Advances in Communications and Media Research

Volume 3

Advances in Communications and Media Research

Advances in Communications and Media Research Volume 3
Patula E. Grossmark (Editor)
2007. ISBN 1-60021-188-7

Advances in Communications and Media Research Volume 2
Anthony V. Stavros (Editor)
2003. ISBN 1-59033-613-5

Advances in Communications and Media Research Volume 1
Anthony V. Stavros (Editor)
2002. ISBN 1-59033-215-6

ADVANCES IN COMMUNICATIONS AND MEDIA RESEARCH

VOLUME 3

PATULA E. GROSSMARK
EDITOR

Nova Science Publishers, Inc.
New York

For permission to use material from this book please contact us:
Telephone 631-231-7269; Fax 631-231-8175
Web Site: http://www.novapublishers.com

LIBRARY OF CONGRESS CATALOGING-IN-PUBLICATION DATA
Available upon request

ISBN 13 978-1-60021-188-1
ISBN 10 1-60021-188-7

Published by Nova Science Publishers, Inc. ✢ New York

CONTENTS

PREFACE

In a society predicated on information, the media has a pervasive presence. From government policy to leisure television, the information age touches us all. The papers collected in this book constitute some of today's leading analyses of the information industry. Together, these essays represent a needed foundation for understanding the present state and future development of the mass media. Current trends in communications as well as media impact on public opinion are studied and reported on.

Predicated on the assumptions that newspapers compete in two markets, (1) for circulation and (2) for advertising dollars, and that segmentation to the local market is crucial to a newspaper's survival, Rosse's (1975) seminal analysis of the newspaper industry resulted in what he called the "layer" (pp. 12-15) or "umbrella" (Rosse and Dertouzos, 1978, p. 166) model of newspaper competition. Rosse's umbrella analogy illustrated oligopolistic behavior of protection and establishment of a sense of order in the industry. Standard economic theory suggests that the dominant firm (s) would exert control by placing an "umbrella" over the entire industry. Minor firms would be regimented to play a clearly defined but subservient part in maximizing industry profits in chapter 1. Each newspaper would operate without competition within its assigned layer of the umbrella, handling the reader and advertiser needs in the geographic area it services. The metropolitan newspaper would encompass and overcover the entire geographic segmentation, and the next smaller size newspaper would provide the same role in its geographic segment.

After 1994, the South African state had available to it few of the usual strategies through which societies construct themselves as nations. Apartheid had compromised language and culture as points of unity, and its association of race with nation had produced a country constituted through a complex of boundaries and cross-cutting ties, which defied any easy or inherited imagining of nation as 'the people'. In chapter 2 is on the basis of this that Thornton argues that the new democratic South Africa must be seen as postmodern (1996: 136-7):

The last decade of the twentieth century saw enormous interest in issues and problems of nation and nationalism, fuelled by competing movements. On the one hand, the breakup of the Soviet Union produced a proliferation of what Anderson (1996: 8) calls "congeries of weak, economically fragile nation-states". At exactly the same moment, globalisation with its transnational systems of production, communication and finance and the attendant routinisation of population displacement and migration, demanded a reappraisal of the role and meaning of nation and nationalism. At the moment of movement from twentieth to twenty first century these issues seem somehow more poignant. Which form of collectivity will best suit the next century? Who will constitute 'we, the people'? Will nationalism and the nation-state continue to serve? National identity has become a major area of concern, particularly as it seems to be portable, no longer irrevocably coupled to territory, but part of a person which goes where he/she goes. In this context, South Africa's nation building quest is fascinating; it is a project to construct the nation as a territorially grounded unity at exactly the same time that globalisation encourages other kinds of collective identities, ones which transcend national boundaries. In such circumstances media in particular can operate paradoxically explained in chapter 3. They can contribute to these other identity processes and simultaneously generate national visions and imaginings which then become available on a global stage, projecting a desired image of South Africa and South Africans outside of the nation's boundaries as well as within.

From its earliest origins, the newspaper presence in the Americas has been adventurous, spirited and, as a matter of history, operating in the midst of often untapped diversity. When British printer Benjamin Harris published his one and only issue of *Publick Occurrences, Both Foreign and Domestick* in 1690, he launched the first newspaper among the English colonies along what is today the east coast of the United States. Sloan and Williams (1994) noted that Harris' journalistic work stood out for its "emphasis on political and dissident Protestant works" (p. 3), reflecting some of the regional politics and British dominance of the emerging states in chapter 4.

There is no official regulation of newspaper ownership in the United States, at either the federal or state level explained in chapter5. Antitrust law, both federal and state, provides a basis for preventing changes in newspaper ownership on a case-by-case basis, but there is no systematic oversight about who or which companies own which newspapers in what markets.

The only two direct laws or policies about newspaper ownership are exceptions to the norm. The Newspaper Preservation Act allows an exception to the antitrust laws (Newspaper Preservation Act, 1970). It allows the owners of competing dailies in the same city to combine business operations, thus creating a monopoly, which would otherwise violate antitrust law. The Federal Communications Commission has prohibited combined ownership of a daily newspaper and a commercial television station in the same market for decades. In 2003 the FCC sought to revise this cross-ownership rule, eliminating it in most cities, but after extensive hearings, a challenge in court, and a partial reversal by a Federal appeals court, it abandoned the proposed changes in early 2005.

The traditional publishing markets are undergoing a fundamental transformation. These developments in chapter 6 pose threats to the traditional newspaper publishing companies, but also create new opportunities for these companies to engage in profitable new ventures and businesses. The companies have to find their way in the turbulence landscape. This study contributes to the existing analysis on the publishing strategy topic by putting the forces active in the market into perspective and by narrowing the problem down to such an extent that current and future changes in the publishing industry can be linked to possible strategic choices. A newspaper publishing company's competitive advantage for copying with the rapid changes in the landscape is not in its current distinctive competencies, but in those that it can grow tomorrow.

The economic health of weekly newspapers, unlike their daily counterparts, appears good in chapter 7. Weeklies are increasing in number and circulation as they enter the millennium (NAA, 2004). The number of dailies, and their circulation, is declining (NAA, 2004).[2]

Weeklies enjoy success despite a lack of professional respect. Scholars also give weeklies less attention than dailies. However, the lack of interest has not diminished weeklies' ability to operate in markets throughout the United States.

[2] Daily circulation trends are complex (America, 2004). Total weekday circulation increased from 58,881,746 in 1960 to 62,766,232 in 1985. Circulation then declined, reaching 55,185, 351 in 2003.

In: Advances in Communications and Media Research, Vol. 3 ISBN 1-60021-188-7
Editor: Patula E. Grossmark, pp. 1-21

Chapter 1

ROSSE'S MODEL AND THE RING MODEL: EXPLAINING NEWSPAPER COMPETITION IN THE MILLENNIUM[1]

Janet A. Bridges[1], Barry R. Litman[2], and Lamar W. Bridges[3]
[1]Sam Houston State University, Huntsville, Texas, USA
[2]Michigan State University, Ann Arbor, Michigan, USA
[3]Texas A and M University – Commerce, College Station, Texas, USA

Predicated on the assumptions that newspapers compete in two markets, (1) for circulation and (2) for advertising dollars, and that segmentation to the local market is crucial to a newspaper's survival, Rosse's (1975) seminal analysis of the newspaper industry resulted in what he called the "layer" (pp. 12-15) or "umbrella" (Rosse and Dertouzos, 1978, p. 166) model of newspaper competition. Rosse's umbrella analogy illustrated oligopolistic behavior of protection and establishment of a sense of order in the industry. Standard economic theory suggests that the dominant firm (s) would exert control by placing an "umbrella" over the entire industry. Minor firms would be regimented to play a clearly defined but subservient part in maximizing industry profits. Each newspaper would operate without competition within its assigned layer of the umbrella, handling the reader and advertiser needs in the geographic area it services. The metropolitan newspaper would encompass and overcover the entire geographic segmentation, and the next smaller size newspaper would provide the same role in its geographic segment.

Rosse (1980) had suggested that "per page per subscriber" (p. 68) economies of scale associated with news production -- the ability to recover "first copy expenses" of producing the story itself, initial costs to begin the reproduction process, maintain "published space," and set up distribution, plus the ability to maintain competitive unit costs as circulation and newspaper size increase -- are more favorable for larger newspapers and effectively preclude

[1]This chapter updates "Rosse's Model Revisited: Moving to Concentric Circles to Explain Newspaper Competition," published in *Journal of Media Economics* 15:1 (2002), 3-19. The update retains as much of the original article as possible, making changes within that context. Where appropriate, material from the original article has been changed to past tense. Portions of the chapter included from the original report are reprinted with permission of Lawrence Erlbaum Associates.

two or more newspapers' ability to compete for the same circulation. These economies of scale have resulted in monopolistic newspaper communities where one organization can more efficiently produce the news product and thus force competitors to either redirect their efforts or abandon the news business entirely (Rosse and Dertouzos, 1978). Under this model, a newspaper's ability to compete outside of its geographic community for circulation -- and by extension for advertising dollars -- in proximate metropolitan areas is restricted by larger organizations that have the ability to take advantage of greater economies of scale.

Rosse explained that a competing newspaper can survive in a given geographic area only if it provides a unique, unsubstitutable product, local information. Therefore, each newspaper's ability to compete in nearby similarly sized or smaller communities will be restricted by organizations that are providing both local news and local advertising to that community, and the distinguishing feature of a newspaper firm is its "locational identity" (Rosse and Dertouzos, 1978, p. 62). Local competition for both audience and advertising come from other local media and advertising outlets rather than from other local newspapers.

Inter-newspaper competition in a given geographic region exists, Rosse (1975; Rosse and Dertouzos, 1978) suggested, among four layers that subordinate both different-sized communities and the newspaper organizations within these communities to newspapers in larger geographic areas proximate to the smaller local environment. Under this model, each larger newspaper disseminates news in a wide geographic area and therefore competes for advertisers beyond the local market. Advantage from economies of scale permits the larger newspaper to competitively restrict or "shade" the smaller newspapers in its geographic area, analogous to sets of differently sized umbrellas. Specifically, a small, local *suburban* newspaper is directly in the competitive shadow of a geographically proximate larger newspaper from a more urban *satellite* community, and both in turn are directly in the competitive shadow of a proximate larger *urban* newspaper that has the capability to provide national and regional news as well as to attract national and regional advertisers. The model does not permit competition within the layers. The inter-layer competition demonstrates a linear pattern, a direct competitive path from the smaller to the larger newspapers that is broken only at a fourth level of small *specialty* newspapers that may circulate within communities with sufficient size and interest in their content.

This direct line from one larger community to the smaller communities within its geographic scope has been displayed graphically as if the industry organizations were located on spokes in an umbrella (see Owen, 1975; Picard, 1989). Although the high absolute circulation of the layer-one newspaper actually represents a smaller proportion of the available market than found in smaller communities, this large, broad circulation base, Rosse (1975; Rosse and Dertouzos, 1978) suggested, also makes the layer-one metropolitan daily attractive to national and regional advertisers. Rosse suggested that the layer-one daily doesn't cover local news beyond the metropolitan area very well, creating an opportunity for locally based dailies to survive.

The Owen (1975) and Picard (1989) graphic models (see also Picard and Brody, 1997) show each lower layer newspaper in a path directly below successively larger papers specific to the geographic area. As illustrated with hypothetical daily newspapers in Figure 1, this direct path demonstrates a linear relationship from each suburban daily through the relevant satellite to the overreaching metropolitan paper.

In Figure 1, each satellite city newspaper is subordinate to and competes with the proximate metropolitan central city daily(ies). These satellite-city newspapers in turn subordinate and compete with a third layer of proximate daily newspapers, the suburban press. The suburban papers proximate to and competitive with satellite-city #2 do not compete with satellite-city #1.

Prior graphic models imply that the fourth, specialized layer is also exclusively linear to the suburban layer, which is appropriate for those publications that focus on a geographic area. But, as Rosse and Dertouzos (1978) indicated, special-interest and weekly newspapers may serve overlapping geographic areas or may circulate within larger communities already served by metropolitan or satellite newspapers.

Metro	-----Satellite #1	-----Suburban #1a	-----Specialized #1a1
			-----Specialized #1a2
			----- Specialized #1a3
		-----Suburban #1b	----- Specialized #1b1
			----- Specialized #1b2
	----- Satellite #2	-----Suburban #2a	----- Specialized #2a1
			----- Specialized #2a2
			----- Specialized #2a3
			----- Specialized #2a3
		-----Suburban #2b	----- Specialized #2b1
			----- Specialized #2b2

Figure 1. A Demonstration of the Linear Interpretation of Rosse's Umbrella Model.

Rosse and Dertouzos (1978) illustrated the umbrella model using circulation penetration in a major, "media-rich metropolitan" (Rosse, 1975, p. 12; Rosse and Dertouzos, 1978, p. 149) U.S. market, San Francisco Bay. Their examination of the New York City market supported the concept of market segmentation within a reconfigured umbrella framework..

Tests of the Umbrella Model

Since Rosse presented his umbrella model, researchers have attempted to replicate Rosse's results in other geographic areas or have used his model as a basis for additional research. Tests of the model have been circulation based, advertising based, and a combination. Rosse (Rosse and Dertouzos, 1978) acknowledged that advertising and circulation are different but said that because of the close ties between the two he did not separate them in his discussions. He said that changes in individual newspaper situations could change advertising markets, and Bridges, Litman and Bridges (2000) argued in the original report that changes in the industry require that each be discussed separately. Rosse also predicted that the larger newspapers would produce better quality products. Because predictions about quality were not the focus, studies addressing content quality and newspaper size were not presented.

Those few studies that had directly tested the linear interpretation of the umbrella model indicated that the coastal situations used by Rosse to support the umbrella model are not necessarily replicated in the rest of the country. Circulation data indicated that the model overall was supported in California (Tillinghast, 1988), not supported in Boston (Devey, 1980), and mixed in Michigan (Lacy and Dalmia, 1991) and nationally (Lacy and Davenport, 1994).

Underlying the umbrella model are at least eight expectations (Rosse, 1975; Rosse and Dertouzos, 1978); some are summaries, some are predictions, and some are assumptions:

1. A summary: Four competitive layers explain newspaper competition in metropolitan areas of the United States. Four additional layers have been identified (Lacy, 1988; Lacy and Dalmia, 1991; Lacy and Davenport, 1994).
2. A summary: Circulation penetration will be denser for lower level newspapers. This expectation was supported when tested in California (Tillinghast, 1988) and partially supported by Bridges, Litman and Bridges (2000).
3. A summary: Circulation penetration for an individual newspaper will be densest in the newspaper's home community. Tillinghast's (1988) figures in California supported this penetration pattern, as did figures in our original report (Bridges, Litman and Bridges, 2000).
4. A summary: Newspapers compete with other layers, not within their own layer. The assumption of no intra-layer competition had decidedly mixed support (Bridges, Litman and Bridges, 2000; Devey, 1980; Niebauer, Lacy, Bernstein and Lau, 1988; Tillinghast, 1988; Lacy and Davenport, 1994; Lacy and Dalmia, 1991).
5. A prediction: If higher-level newspapers are forced to pull back their geographic reach, the lower-level newspapers will expand their circulation base. Tillinghast's (1988) findings with a dominant metropolitan daily showed little effect; Devey (1980) provided indirect support; Niebauer et al. (1988) concluded metropolitan intracity status did not affect the suburban market.
6. An assumption: As newspapers move farther from layer one, scale economies restrict a smaller newspaper to content that is increasingly local and of lesser "quality." The assumption that profitable newspapers will commit more dollars to editorial quality than less profitable newspapers has been discussed elsewhere as "financial commitment" (see Lacy, 1989; Lacy, 1992; Lacy and Fico, 1989; Litman and Bridges, 1986; Rosenstiel and Mitchell, 2004). As noted earlier, content quality was not part of the original discussion and will not be in this chapter. However, we note that the 2004 report from The Project for Excellence in Journalism (2004c) indicated that increases in newspaper profit margins have not been reflected in newsroom budgets.
7. An assumption: Penetration analysis, an analysis of the circulation market, can serve as a proxy for an analysis of advertising markets. Lacy's (1984) regional study separated circulation and advertising effects of competition. Also see Bridges, Litman and Bridges (2000) and Lacy (1985).

8. An assumption: As newspapers move farther from layer one, advertising will become increasingly local. Evidence is indirect. National advertising figures (Busterna, 1987) refuted both Rosse (1975) and Owen (1975); Tillinghast (1988) noted collective suburban sales competed with the metropolitan daily; Lacy (1988) had suggested that grouped weeklies can compete with metropolitan dailies; Lacy and Sohn (1990) found differences in substitutability of advertising and news content in Detroit and Denver.

ROSSE'S MODEL AND NEWSPAPERS AT THE MILLENNIUM

The pre-2000 tests of the umbrella model raised concerns about both the nature of intra-layer competition and the use of circulation as a proxy for advertising data. Circulation and/or penetration figures generally supported the verbal -- but not later graphic -- descriptions of the model, even though every study identified anomalies in the circulation patterns. Rosse's conclusion that the local community is the most effective source of circulation for a daily newspaper was not disputed by the Bridges, Litman and Bridges (2000) ring model. What the ring model emphasized is that the local community is no longer the only source of circulation. The industry has been concentrating on ways of maintaining relevance in the local community (see C. S. Stepp, 1999), although at the same time, larger newspapers are continuing to attempt to maintain attractiveness in smaller surrounding communities. The rigid, linear interpretation that restricts competitive relationships no longer seemed to describe the newspaper industry.

Rosse's model was created in a vastly different newspaper industry. The major metropolitan daily was the dominant newspaper firm in the state or region, and, because of their inability to obtain the economic efficiencies of larger metropolitan competitors, the other dailies, weeklies, and specialty newspapers were in clearly inferior and weakened positions. As Rosse (1980) predicted, telecommunication and production technology have provided more equitable distribution of economies of scale, access to information and readers, and incentives for possible intra-industry cooperation. Rosse's use of circulation as the indicator of competition was logical but too narrow for today's newspaper industry.

As was demonstrated by Bridges, Litman and Bridges (2002), the newspaper industry structure at the millennium does not fit either (a) the exclusivity (or linearity) of the inter-layer competitive relationships, (b) the assumption that newspapers in the same layer will not compete, or (c) the exclusive use of circulation to test the model. Many of the industry changes at the end of the 20th Century had been anticipated by Rosse.

Exclusivity (or Linearity) of the Inter-Layer Competitive Relationships; Newspapers in the Same Layer Will not Compete

Rosse (1975) had predicted that if lower-layer newspapers could expand their news base, they would compete more intensely with the upper-level dailies. The industry was and still is experiencing efforts by both satellite and suburban papers to circulate beyond their geographic boundaries, regardless of direction of this competition. And, as a result of conglomerate ownership and technological advances, these newspapers were more efficient if they wished to provide news beyond the local environment to their readers (see Bass, 1999). The trend toward converging media and the availability of web technology made this expansion even more feasible. Suburban newspapers are also attempting to attract commuters from other geographic areas, because the workforce is commuting to locations other than the metropolitan central city and is useful to the advertisers in the work communities ("Single Copy," 1996; H. Stepp, 1996). The mobile work force alone suggests intra-level competition among suburban newspapers. Special, zoned editions and the Internet are providing additional opportunities for newspapers to circulate beyond their communities. The Newspaper Association of America (2004) indicated that individuals who report workday use of newspaper web sites also report the newspaper sites as their most frequently used media sites. Conniff (1995) had suggested that important topics also become "local."

Rosse used the competitive San Francisco Bay area to demonstrate the umbrella model in the 1970s. Bridges, Litman and Bridges (2002) attempted to find a similar competitive situation in a different geographic area where satellite newspapers operated and located competing large dailies in the Tampa, Florida, region. Audit Bureau of Circulations (ABC) data (1996) indicated that the Tampa region had two large daily newspapers, the *Tampa Tribune* and the *St. Petersburg Times*, competing against each other. Tampa is located in Hillsborough county and St. Petersburg is in Pinellas county. The *Times* maintained a separate SRDS (1995) listing under Tampa as well as under St. Petersburg. No finite definition separates metropolitan daily newspapers from satellite newspapers, but the size of these communities compared to communities of metropolitan daily newspapers such as *The Dallas Morning News*, the *San Francisco Chronicle*, or even the *Atlanta Journal/Constitution* led them to categorize the two Florida dailies as smaller, satellite dailies. The *Miami Herald*, which circulated in Hillsborough county, provided weak metropolitan "shade." The *Orlando Sentinel* was also a smaller presence in Hillsborough county.

As Lacy (1988) suggested, a national layer also existed. *USA Today* and the *Chicago Tribune* circulated in both Hillsborough and Pinellas counties, creating a separate national layer. Thus, as defined, the layers in the Tampa area were national, metropolitan, satellite, suburban, and weekly. In this market suburban dailies demonstrated multidirectional competition.

Raw penetration percentages supported the linear interpretation of Rosse's umbrella model. But a closer look indicated that the *St. Petersburg Times* was making inroads into the Tampa city circulation, an example of intra-layer competition in the satellite layer. The suburban layer also suggested a second interpretation. With one exception density figures supported Rosse's belief that smaller dailies will have higher penetration in their home communities than larger dailies will in theirs. The density figures also supported intra-layer competition for circulation.

ABC analysis of county circulation for daily newspapers in the same Florida market showed the *Tampa Tribune* increased its penetration in Citrus County (where Crystal River is located); there was a minor amount of circulation from the suburban dailies in the satellite layer, but overall the patterns of circulation competition did not change. County penetration was consistent with city penetration.

The Circular Ring Model

Although the Florida figures were not definitive, combined with prior studies they suggested that the three daily-newspaper layers identified by Rosse still exist, but they do not function in the linear fashion illustrated by graphic models. Local circulation density patterns predominantly support Rosse's model, but there are deviations from Rosse's expectations in the paper editions. In today's market the nature of newspaper competition, especially for advertising, is multidirectional rather than inter layer. The area of natural advantage for each newspaper will depend not only on its geographic location but also on the role the newspaper has chosen to play in its particular market at any particular time.

Bridges, Litman and Bridges (2002) proposed that this multidirectional competition can be graphically displayed, not as rigid spokes in an umbrella, but as layers of concentric circles that preserve much of the hierarchy proposed by Rosse but accommodate the behavior of the newspaper rather than just its location.

As Figure 2 demonstrates, for a given metropolitan area the circular, or ring, model of competition shows four concentric circles and an ellipse.[2] Three-dimensionally, the circles should be perceived as different-sized disks layered on a central post. Each disk represents a different-sized newspaper type; the center would be the center of the metropolitan area. The unlimited possible movements of these disks in relationship to each other illustrates the variety of competitive relationships among newspapers in a market that is expected to vary both for circulation and for advertising.

Underlying the metropolitan area is a layer of national or regional newspapers, which, except for their home base, do not generally originate in the metropolitan market (see Lacy, 1988). Because the national/regional layer underlies more than one metropolitan area, these newspapers are represented by a broken or porous boundary that indicates circulation extends beyond the metropolitan area illustrated. This layer may include a major state newspaper(s) or national newspapers, such as *USA Today*, *The New York Times*, *The Wall Street Journal,* and the *Christian Science Monitor*. Nondaily newspapers may also circulate at this level. Also included in this layer would be newspapers that circulate in a broad geographic area of the country. Circulation density in the metropolitan market should be weak, but competition from this broad-based layer would be with any level of that market.

The small ring in the center encloses a circle that represents the densest circulation area of the metropolitan paper. The ability of both smaller and larger newspapers to compete in the metropolitan area, especially for advertising, will vary geographically and is recognized by the ellipse marked "overlap." Consistent with Rosse, this geographic center area is where the metropolitan newspaper should have a natural advantage. Without a geographic presence in the Tampa/St. Petersburg market, the Florida metropolitan newspapers, for example, had weak circulation in the suburban and satellite layers.

[2]Lacy and Dalmia (1991) cited an unpublished paper by S. Dalmia, "Ring Theory as an Alternative to the Umbrella Model of Newspaper Competition: A Study" but did not expound on the information. The paper, a derivative of unpublished work of Barry Litman begun in 1985, was unavailable.

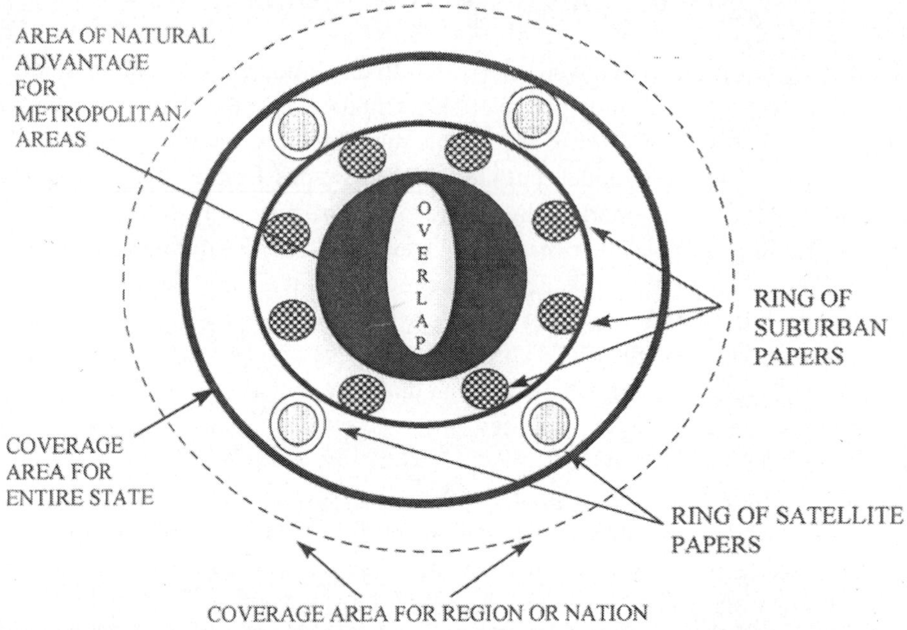

Figure 2. Concentric Circles: The Ring Model of Newspaper Competition. Originally appeared in "Rosse's Model Revisited: Moving to Concentric Circles to Explain Newspaper Competition" by Bridges, Litman and Bridges, *Journal of Media Economics* 15(1), 3-19. Reprinted with permission from Erlbaum.

The next circle is dominated by a ring of suburban papers, which Rosse had subordinated to proximate satellite dailies. However, prior studies have indicated that in some areas a satellite daily may not be a factor for suburban newspapers (see Devey, 1980; Lacy and Dalmia, 1991; Lacy and Davenport, 1994). Notable, then, is the change from Rosse's model that had associated each suburban paper with one particular satellite daily. Suburban newspapers can be located in suburbs that are geographically proximate to and in competition with either metropolitan or satellite communities, or both; and inter-layer competition may come from the metropolitan newspaper, through either zoned editions or regular content, or from satellite newspapers. Technology now allows large newspapers to provide good, timely coverage of smaller communities where they have an interest, and chain affiliations can provide news beyond the local area to smaller newspapers wishing to compete at that level. Chain resources can expand geographic competitive ability of the smaller newspapers, especially for advertising packages.

The third circle represents the satellite newspapers, which are larger than the suburban newspapers. This circle also suggests that the combination of suburban, satellite and metropolitan newspapers can provide statewide coverage.

Weekly newspapers, shoppers, ethnic and other specialized nondaily newspapers operate within any geographical area (see Rosse, 1975). We are suggesting these newspapers do not present new layers, but they behave uniquely within the circular rings. Weeklies are being purchased by larger chains and are operating competitively (see Bissinger, 1999). Ethnic media are primarily local; however, ethnic media are merging (Project for Excellence, 2004f).

In the daily press, some zoned editions behave as pseudo weeklies. Two major papers have experimented with free newspapers directed toward the 18-to-34-year-old market, and the industry is developing special Spanish-Language editions to reach the growing Hispanic population (Morton, 2004a; Project for Excellence in Journalism, 2004a). Consolidation in Hispanic news properties is expected to result in a national presence for recently formed Impremedia (Kim, 2004). The language used to reach the Hispanic market, especially the young market, may see change, as 45 percent of the 18-29 year old Hispanics consider themselves "largely English speakers" (Solís, 2004, p. 6D).

Exclusive Use of Circulation to Test the Model

Economic competition among newspapers today cannot be discussed without a separate discussion of advertising practices. The audience is a separate "product" produced by newspapers. This product is marketed to advertisers (see Picard, 2004), and the substitutability of that product is different than it was in the 1970s.

As noted, Rosse acknowledged advertising as a distinct market; he also said circulation is not the major source of newspaper revenue. Up to about 80 percent of a newspaper's incomecan be attributed to advertising (Project for Excellence in Journalism, 2004a; Picard, 2004; Maynard, 1996; Picard, 1993; Rosse, 1975; Rosse and Dertouzos, 1978), and based on citation dates this percentage has been consistent across time. Rosse tied local circulation to advertising revenue. Newspaper advertising revenue is certainly dependent on circulation, but today's newspaper owners are working outside local circulation parameters to attract advertising revenue. Since the late 1970s when Rosse presented his umbrella model, the newspaper industry has focused seriously on expanding its advertising penetration beyond the local areas and on making smaller daily, and even weekly, newspapers attractive to nonlocal advertisers. The local market is weak; retailers have been consumed by large chains, and the large chains want more than a small local market for their advertising dollars (Morton, 1998).

The current trend includes both cooperative industry efforts and extensions of group ownership that combine geographically proximate suburban newspapers, both daily and weekly, into one package for national and regional advertising. This regional package competes with the other levels of newspapers. Rosse (1975; Rosse and Dertouzos, 1978) had said that changes in advertising prices could shift advertising markets into a different layer. Today's industry reflects these shifts.

Below is an updated description of efforts by the newspaper industry to compete in the national advertising market and of ownership trends that move the local newspaper into a broader market for advertising. The Bridges, Litman and Bridges (2000) report presented one example of cost-benefits. The millennium newspaper rate card is completely open to negotiation (Morton, 2004b, and Efchak, 2004), online and print classifieds are frequently sold as a package (Project for Excellence, 2004e) and, therefore, speculation would be inappropriate.

Industry Cooperative Efforts

The first major cooperative industry effort was in the early 1980s when the newspaper industry adopted the Standard Advertising Unit (SAU), a system that signaled a cooperative attitude among competing daily newspapers, regardless of their placement in the umbrella model. An accommodation to potential advertisers, the SAU signaled that newspapers were committing to standardizing their page formats to six columns and to standardizing their advertising sizes. Under the SAU format, an ad of one size would fit the same way into all newspapers adopting the SAUs, thus reducing first-copy costs for advertisers who might wish to advertise in smaller newspapers or in several members of one chain (Conover, 1985).

The NAA more recently proposed a standard that will decrease web width, resulting in smaller page sizes for newspapers. The assumption is that newsprint costs will be reduced (Web Width, 2000). Ad sizes will be smaller for newspapers adopting the new page sizes, but the NAA has not proposed new SAU formats. Rather, the association acknowledged that publishers may decide to follow PASS, meaning use their "preferred ad-submission sizes" (paragraph 7 or Hernandez, 2000, paragraph 4). Picard (2004) noted that newsprint accounts for about half of a newspaper's production/reproduction costs, which make up 30 to 35 percent of the medium's operating expenses, and the Project for Excellence in Journalism (2004b) reported that newsprint prices rose in 2003. Publishers have already used contentchanges to reduce the amount of paper consumed (Callahan, 2002; Rose, 2002).

About a decade after the SAUs, through the Newspaper National Network[3] (NNN) the industry began cooperating nationally in seven common retail categories to place ads in combinations of their member daily newspapers which circulate at all levels of the "umbrella" (Albers, 1997; Bogart, 1998; "Newspaper National," 1998). According to the NNN web site, nearly 9,000 newspapers currently cooperate with the organization (Newspaper National, 2004a) and the cooperative advertisingplacement has expanded to "all advertising categories where newspapers receive less than 10% of total media spending" (Newspaper National, 2004b). The organization is now working with daily newspapers to address the use of ad placement in individual sections such as business or sports (Fine, 2004). The newspaper industryhas also begun to cooperate in on-line classifieds, once a mainstay of local newspapers. On-line capability is being used for real estate and automotive advertising, for example. One consortium, Classified Ventures, brings together 170 web sites for newspapers and television (Project for Excellencein Journalism, 2004b). There was also some attempt to reproduce coupon clipping on line (Hilts, 1997; Maynard, 1996; Miller, 1998; Outing, 1998; "Tracking," 1996), and two media groups financed a "social networking site" (OPA, 2003, graph 1).

Combining Geographically Proximate Newspapers: Clustering

More important to this discussion of market structure is the regional focus of the suburban "groups." Newspaper analyst Morton (1997) called this proliferation of regional groups "clustering," a term he said would be the buzzword of the 1990s. The industryhas

[3]The original Bridges, Litman and Bridges (2000) reportincorrectly reported this organizationas the National Newspaper Network

always had groups of smaller newspapers, but as Morton explained, groups are now buying and trading newspapers with regionalism as the goal (see also Moore, 1999; Ridder, 1998). Recent acquisitionsindicated that these clustersare including both daily and weekly newspapers, and Wrubel (1998) suggested that magazines and other "niche" publications may become part of these clusters (see also Potter, 1998). Picard (2004) said newspaper advertisers are trying to reach a mass audience rather than specific segmented audiences. As a group's clustering becomes more and more regional, an advertiser can reach an entire geographic area with one ad purchase. Consolidated production and administrative activities facilitated by clustering produce the potential for production economies of scale that were available only to the larger newspapers in Rosse's analysis (see Martin, 2003). These economies can reduce costsand enable the group to present an advertising package to retailers whose buying decisions have become more and more national and regional. However, the Lacy and Martin (2004) review indicated that clustering behavior has increased advertising rates. The authors did not specify, but based on prior studies, we assume that published rates were the criterion.

What these trends imply is that, at least at the suburban level, competition for advertising across layers is resulting in cooperative ventures within layers beyond the local marketThis cooperation creates overlap in the pool of potential advertisers, a contradiction of the assumption behind the linear interpretation of the umbrella model. The inter-layer competition that is the end result of this cooperation will make the regional package competitiveas a unit rather than on a one-on-one basis with larger newspapers. Because deliveryand newsprint costs still restrict indiscriminate circulationgrowth, demographics rather than density may dictate competitive status.

Competition for advertising becomes not an attempt to eliminate another suburban newspaper but instead a focus on the ways daily newspapers can cooperate to become more attractive for advertisers outside of the local marketBy responding to these advertiser needs, the suburban newspapers are moving away from the linear umbrella relationships and becoming a fluid, changing layer that competes differently under differing regional conditions. As advertising becomes a major revenue component of newspaper web sites, the pattern of competitionwill change again (see below).

NEWSPAPERS ON LINE

The NAA reported that as of April 2003 almost 1,500 daily newspapersin North Americahad web sites. With 1469 U.S. daily newspapers and 102 in Canada (Newspaper Association of America, 2003b), we can only conclude that nearly every daily newspaper has developed a web site; Picard (2004) set the number at 90 percent. Greer and Mensing (2004) located 1279 U. S. daily newspaper sites in 2003.

As newspapersmove through the first decade of the millennium, competition within the newspaper industryis changing to accommodate both on-line access to competing newspapers and convergencewith other media formats. For example, a different type of clustering behavior has been labeled "partnerships," and involves nonacquisition, multimedia sharing (Ahrens, 2003; also see Associated Press, 2003, for Iraqi warcooperation; Demo, Spillman and Dailey, 2004) by "most major news organizations" (Ahrens, 13). Newspapers are sharing news with other newspapers, with television stationswith cable stations, and with on-line sites, which do not belong to the overreaching owner organization. Demo, Spillman and Dailey (2004) said nearly 30 percent of U. S. daily newspapers partner with television stations. Some of these partnerships are international and most extend the newspaper's visibility geographically. Runett (2004) said the "cross-media partnerships" (p. 10) begin with print reporters sharing with the newspaper's 24-hour website. Ahrens (2003) suggested advertising rates and brand-awareness considerations underlie the sharing behavior. He said the partnered Internet sites are expected to move viewers between sites, helping both partners, but he found no definitive results. In a cross-ownership situation, a representative from the *Chicago Tribune* said the paper felt that its readershipincreased after *Tribune* reporters presented an investigative story on WGN-TV (Simpson, 2003). The Project for Excellence in Journalism (2004d) predicted that online growth will lead to direct competition between media from what traditionally have been different fields.

On-line editions may be reinforcing the local focus of at least some newspapers, consistent with Rosse (1975), as the on-line presence of most newspapers attracts individuals interested in local news and local advertising (Coats, 2002). The newspaper site is the "top internetsource for local news and information" for 83 percent of newspaper web-site users (NAA, 2003a), and newspaper on-line sites are "the most visited local sites," more so than television (Picard, 2004, p. 118; see also Francese, 2003). A 2002 report in *presstime* showed that, not surprisingly, newspapers use their sites for businesspurposes as well as for news (Rabasca, 2002). The newspapers are building databases of registrants, targeting both advertising and news, and providing the expected access to subscriptions, address changes, bill payment and reports of delivery problems. Less common, some are charging for on-line subscriptions, especially when the subscriber is not a print subscriber, and others provide limited access but charge for depth information. Some papers are charging regardless of print-subscription status, and some are charging for specific information. There seems to be no universal behavior as newspapers look for the possibility of up-front revenuefrom the on-line sites (Rabasca; see also Project for Excellence2004b; Greer and Mensing, 2004).

Outing's (2003) interviews with on-line consultants suggested that newspaperscould use sponsored on-line targeted material, similar to special editions or special sections, a form of niche advertising. Earlier Outing (2002) reported Borrell's suggestion that for larger newspapers with expensive ad rates, online editions could serve smaller businesses that need a niche or businesses that want to reach the younger readers attracted to on-line editions.

Picard (2004) noted that " . . . very few papers have found ways to make these operations profitable" (pp. 117-8). However that assessment may be changing, at least for some newspapersBorrell (2004b) reported that newspaper revenueis expected to pass $1 billion for 2004, and (2004a) 83 percent of the newspaper sites reported profitability for the year, averaging about 60 percent profit margin. Their data (2004a) were based on based on information from 711 newspaper and televisionweb sites. However, the Project for Excellence in Journalism (2004e, 2004g) cautioned that start-up costscost accountingand cannibalizing from the print editions affect profitsBorrell (2004a) said newspapers were reducing their classified advertising "dependence" (p. 1), but later (2004b) said the industry was still "over-dependent" on the classifieds. In 2002 emphasis was on local advertising (Borrell Associates, 2003). In 2003 revenue for on-line job-related ads dropped as hiring declined, and predictions about future revenue from on-line advertising was mixed (Gough, 2003). Competition comes from outside the newspaper industryas well as from inside. For example, competition for job-related ads, especially in classifieds, comes from major nonnews sites such as Monster.com (Project for Excellence in Journalism, 2004b), and eBay and city-matching web sites have been drawing merchandise-sales classifieds that attract readers (Outing, 2004; Mangalindan, 2004). Nonnews sites, notably AutoTrader and Homestore, have alliances with on-line networks and provided classified competition in automotive and nonspecific merchandise, and Borrell predicted that the Yellow Pages will grow as local competitors. Local television and direct mailwere not major factors in 2002 local on-line advertising (Borrell, 2003) but local television websitesare becoming increasingly competitive (Borrell, 2004b).

Technology is making classified advertisingeasier and more competitive. IPIX, which had provided graphics for e-Bay (Morrill, 2004), is marketing AdMission Classifieds, a graphics program designed to allow "rich mediafor classified advertisers, including photos, slide shows, online editing and other options during on-line or telephoneconversations with newspaper sales representatives (IPIX AdMission, 2004). According to Mandese (2003), DoubleClick concluded that rich media would become the "online advertising standard" (1). The IPIX program has a search feature (Morrill, 2004; Rodgers, 2004) that can be targeted by geographic density (Rodgers, 2004), and is expected to handle both print and on-line ads in one effort (IPIX AdMission, 2004).

Measurement of on-line traffic is also an issue (Bialik, 2004; Hansell, 2003; O'Connell, 2001; NAA, 2004; Stone, 2002), both for a publication count in general and for audiences for specific contentespecially ads. The Audit Bureau of Circulations (ABC) as of November 2003 permits publishersto count electronic editions separately when subscription charges are 25 percent of the print price (Audit Bureau, 2004; see Rose, 2003, for an example). Friedman's (2003) essay reported that ABC counted digital editions that replicate a newspaper's print edition exactly. However, ABC may be a little broader, considering a paid count when the electronic edition is "consistent in content with the print version" (Audit Bureau, 2004, page 1). The ABC offered to review electronic editions to determine compliance (Audit Bureau, 2004). The ABC apparently has also accepted researchin Britain and will begin including on-line figures as part of the British press reports ("Digital publications," 2003). Nielsen has been monitoringon-line traffic in general, although some publications have questioned the Nielsen figures and use of software (Elkin, 2003). For a discussion of the measurement issuessee Stone (2002). In contrast to the newspaper mass audience tradition, Borrell Associates found that on-line advertisers were interested in targeted information rather than global page hits or lists of "unique users" (Borrell, 2002, p. 2). However, Borrell found no consistent measurement being used by these advertisers. Nor was any measurement being used to assess the actual impact of online advertisingAs Borrell indicated, at this point newspapers have an opportunity to advise their advertisers.

Moving back to the original Florida sample used by Bridges, Litman and Bridges (2002), all but one of the regional daily newspapershas an active web site, and--based on the functioning sites on January 6 2004--these sites follow the dailies' earlier print competitive behavior, competing within as well as across layers. For example the on-line *St. Petersburg Times* shows online "Times" in two regional news sections as well as links to eight separate regions, which include four counties, including Tampa's Hillsborough, and the Tampa Bay area itself. A separate link is set up for Tampa Bay Business, and the 2003 *Editor and Publisher Yearbook* (Maddux, 2003) shows separate managing editors listed for Tampa, as well as for Clearwater and the North Suncoast. The St. Petersburg daily shows the highest circulation in the region (Maddux, 2003).

The *Tampa Tribune* on-line site showed links to four counties, including St. Petersburg's Pinellas, and six communitiesprimarily within the Tampa/Hillsborough area. It also had links to regions within Tampa itself. Both the Tampa and the St. Petersburg on-line dailies showed links to Brandon.

Moving from the satellite layer to the suburban layer, the on-line *Bradenton Herald* showed two links, one to Bradenton news and a second to the Eastern part of Bradenton's Manatee county. The Bradenton site advertised movies, weather and a generic call for volunteering in Sarasota but did not appear to position itself as a competitor of the *Sarasota Herald-Tribune*. The Sarasota daily, however, is more aggressive. The Sarasota on-line edition showed links to eight areas, three to cities in Sarasota county without daily newspapers, two in the *Bradenton Herald's* Manatee county, one to the Tampa Bay area, and one to Charlotte, which abuts the south end of the region. The 2003 *Editor and Publisher Yearbook* (Maddux, 2003) indicated that the Sarasota daily has separate publishers for both the Manatee and the Charlotte/Englewood areas.

The *Sarasota Herald-Tribune*, part of the *New York Times* regional newspaper group, showed the largest circulation in the suburban layer (Maddux, 2003). Its sister publication, the *Lakeland Ledger*, is also part of that group but about 2/3 the size of the Sarasota daily (Maddux, 2003). The *Lakeland Ledger* is a less aggressive competitor. The Lakeland daily showed only two links, and these were within the county. The *Citrus County Chronicle* listed an on-line site, but it was not active.

These Florida newspapers are also showing some broadcast cross ownership which may speed up convergence in the area. Lakeland listed the *New York Times* Company Broadcast group; Sarasota listed SNN6, Sarasota News Now; and Tampa listed WFLA-TV (Maddux, 2003).

CONCLUSION

The Bridges, Litman and Bridges (2002) circular modification of Rosse's model recognized the variation of newspapers in the industry both in size and in geographic location. It preserved Rosse's three daily-newspaper layers, but altered the relationship among them and better illustrated both the variety and the fluidity of market roles and metropolitan geographic relationships in this country. The model permitted movement both within and among the layers. Expanding technology was expected to remove the land-locked constraints on news providers and require a new model of competition.

The greatest change in technology since the original model was presented is the advances in on-line capability. The technology has opened possibilities to the daily newspaper but has not yet reached its potential. Dailies demonstrate no consistent business model, and competition from the on-line perspective, at least within this purposive area, stays within the parameters suggested by the ring model. Those dailies that choose to compete with on-line editions still compete within rings and across rings to smaller markets, regardless of that community's newspaper presence. The larger dailies are more aggressive within their layers than the smaller dailies within the same layer. However, reports from 2003 (Borrell, 2004) indicated that some newspapers were beginning to see a profit from their on-line efforts, and we expect to see expansions of on-line competition and, as suggested earlier, changes within the ring model.

We still have the concern articulated by Bridges, Litman and Bridges (2002), acknowledging the limitations of the purposive sample used for demonstration. But we repeat that purposive sampling of regions has been used consistently in testing the umbrella model, even though criteria and other methodologies have not been consistent. The peninsular Tampa-St. Petersburg region is similar geographically to the San Francisco area used by Rosse originally, which displays geographic uniqueness also noted by Lacy and Dalmia (1991). The linear interpretation of Rosse's model may still operate in some metropolitan areas, and Bridges, Litman and Bridges had suggested that future research should test the circular model in other geographic markets.

As Bridges, Litman and Bridges (2002) concluded, the circular, ring model emphasizes the ability of each newspaper to compete differentially at any level in its broad geographic market, particularly for advertising. Circulation patterns may differ, but trends in advertising, a newspaper's major revenue source, suggested that the linear interpretation of Rosse's model no longer represents the newspaper industry, where the comfortable world of fixed market segmentation no longer exists. The circular, ring model presumes a fluidity of competitionthat responds to changes in either advertising prices and/or circulation, regardless of the local area of natural advantage. This is still a dynamic notion of competition between and among firmsoperating at all layers, as each medium constantly seeks to preserve its domain and product niche against the onslaught of new technologies and new media servicesRosse had predicted many of the changes in the newspaper industry and had articulated scenarios that would result in other changes. This concentric-circle model of newspaper competition preserved most of Rosse's taxonomy while it incorporated the market changes.

Technological capabilities have placed the newspaper market in a state of continued exciting change, and the underlying ring modelis acquiring an overarching dome, a three-dimensional addition stretching the daily newspaper's reach without technological boundaries. Online capability and returnsfrom these online efforts have the market in a state of turbulence, as business models are tried and overtures are made beyond the local market. The only certainty is expectation for continued growthin the online areas, greater sophistication in the technical capabilities, and stronger inter-media cooperation.

REFERENCES

Ahrens, F. (2003, November 6). Media giants getting together: Firms see partnerships as low-cost way to expand reach, coverage. *The Washington post*, p. E01. Retrieved November 15, 2004, from http://web.LexisNexis.com/universe/document?_m=2be92406545

Albers R. R. (1997, December). Newspaper national network: Year three. *presstime, 19*, 29-30.

Associated Press (2003, March 17). 'NY Times,' 'Boston Globe' partner with CNN: Newspaper reporters will help with Iraq coverage. Editor and publisher online. Retrieved April 13, 2003, *fromhttp://editorandpublisher.printthis.clickability.com/pt/cpt?action=cptandexpire=andurlID=5725369andfb+YandpartnerID=60; also available at http://*

www.editorandpublisher.com/editorandpublisher/headlines/article_display.jsp?vnu_content_id=1840367

Audit Bureau of Circulations. Newspaper rule clarifications: Qualifying electronic editions as paid circulation (updated Nov. 2003). Retrieved September 3, 2004, from *http://www.accessabc.com/members1/n_electronic.htm*

Bass, J. (1999, July/August). Newspaper monopoly, *American journalismreview, 21*, 64-86.

Bialik, C. (2004, November 15). Web publishers move toward standardizing ad-viewership data. *The wall street journal*, pp. B1, B6.

Bissinger, B. (1999, December). Feeling the heat. *American journalism review, 21*, 49-65.

Bogart, L. (1998, February). Who's selling national ads? *presstime, 20*, 33.

Borrell Associates Inc. (2002, April). *What advertisers want from online newspaper sites: Online measurementproject*, Phase I. Executive summary. Portsmouth VA: author.

Borrell Associates Inc. (2003, April). What newspaper web sites earn: A bench marking report. *Executive summary*. Portsmouth VA: author.

Borrell Associates Inc. (2004a, May 11). *Newspaper and TV websites now a $1 billion business*. Press release.

Borrell Associates Inc., (2004b, August 3). *Preparing for 2005 budgets*. Memorandum.

Bridges, J. A.; B. R. Litman; and L. W. Bridges (2002). Rosse's model revisited: Moving to concentric circles to explain newspaper competition*The journal of mediaeconomics, 15*, 3-19.

Busterna, J. C. (1987). The cross-elasticity of demandfor national newspaper advertising. *Journalism quarterly, 64*, 346-51.

Callahan, P. (2004, March 18). Endangered species: Newspapers' stock listings: Tiny type eats up newsprint, doesn't sell ads--but is beloved by older readers. *The wall street journal*, pp. B1, B6.

Coats, R. (2002). Power users: A profile of online newspaper consumers. *Digital edge report1*. Vienna VA: Newspaper Association of America.

Conover, T. E. (1985). *Graphic communications today* (pp. 300-301). St. Paul: West Publishing Co.

Demo, L; M. Spillman; and L. Dailey (2004, October 15). *Newsroom partnership survey*: Executive summary. Muncie IN: authors.

Devey, S. M. (1989). Umbrella competitionfor newspaper circulationin the Boston metro area. *Journal of mediaeconomics, 2,* 31-40.

Digital publications to receive ABC verification (2003, December 18). *Mediaweek online*. Retrieved January 8, 2004, from http://www.mediaweek.co.uk/ThisWeek.News.View.aspx?ContentID=11105

Efchak, E. (2004, August 6). Verbal comments during *"The future of newspapersAn executive briefing."* Panel discussion at the nationalmeeting of the Association for Education in Journalism and Mass Communication, Toronto.

Elkin, T. (2003, March 10). Nielsen rolls net-traffic gauge. *Advertising age74*, p. 2 [Electronic version].

Fine, J. (2004, February 9). Papers' ad group goes on offensive: NNN section-specific push targets mags, cable. *Advertising age*p. 6 [Electronic Version]. Retrieved August 28, 2004, from http://www.nnn-naa.com/headlines_adAge.htm

Francese, P. (2003, July/August). Stop the press! Newspapers--still a media force to contend with--can turn demographic declines into gains through digitization, customization and marketing basics. *American demographics25*, pp. 38-39.

Friedman, D. (2003, November 5). Business: From freeto fee in 10 easy steps. *Online journalism review*. Retrieved January 14, 2004, from file:///Macintosh%20HD/Research-3/OJR%20article-convergence

Greer, J., and D. Mensing (2004). U.S. news web sites better, but small papers still lag. *Newspaper research journal, 25*, 2, 98-112.

Hansell, S. (2003, February 24). Ratings agency says it erred in measuring web site use. *The New York times* [Electronic Version]. Retrieved March 6, 2003, from http://www.nytimes.com/2003/02/24/technology/24net.html?pagewanted=printandpositio n=bottom; also available at http://web.LexisNexis.com/universe/printdoc

Hernandez, D. G. (2000, February 8). NAA board approves suggested page sizing for smaller widths: Task force recommendations put forth as guidelinesfor 50-inch web, not new standardsNAA news release. Retrieved August 21 2004 from http://www.naa.org/technologypressweb/50inchPR.pdf

Hilts, E. (1997, December 6). Online coupon strategies for small newspapers. How two papers at opposite ends of the country approach Internetcoupons from different directions. *Editor and publisher, 130*, p. 30.

IPIX AdMission (2004, June 17). AdMission classifieds now available for newspapers: Feature-packed new photo ad solution integrates with minimal effort. Press release. Retrieved June 28, 2004, from http://biz.yahoo.com/prnews/040617/sfth029_1.html

Kim, Q. S. (2004, January 16). Hispanic-newspaper battle shapes up. *The wall street journal*, p. A8.

Lacy, S. (1988). Competing in the suburbs: A research review of intercity newspaper competition. *Newspaper research journal, 9,* 69-76.

Lacy, S. (1984). Competition among metropolitan daily, small daily and weekly newspapers. *Journalism quarterly, 61*, 640-44, 742.

Lacy, S. (1992). The financial commitmentapproach to news media competition. *Journal of media economics, 5*, 5-21.

Lacy, S. (1989). A model of demandfor news: Understanding the impact of competition on daily newspaper content. *Journalism quarterly, 66*, 40-48.

Lacy, S. (1985). Monopoly metropolitan dailiesand inter-city competition. *Journalism quarterly, 62*, 640-44.

Lacy, S. and S. Dalmia (1991). *Michigan newspaper competition from 1980 to 1986: Expanding the geographic application of the umbrella model.* Paper presented at the annual meeting of the Association for Education in Journalism and Mass Communication, Boston, MA.

Lacy, S. and L. Davenport (1994). Daily newspaper market structure, concentration and competition. *The journal of mediaeconomics, 7*, 33-46.

Lacy, S. and F. Fico (1989). *Financial commitment, newspaper quality and circulation: Testing an economic model of direct newspaper competition.* Paper presented at the annual meeting of the Association for Education in Journalism and Mass Communication, Washington, DC.

Lacy, S. and H. J. Martin (2004). Competition, circulation and advertising. *Newspaper research journal, 25*, 1, 18-39.

Lacy, S. and A. B. Sohn (1990). Correlations of newspaper contentwith circulationin the suburbs: A case study. *Journalism quarterly, 67*, 785-93.

Litman, B. R. and J. Bridges (1986). An economic analysis of daily newspaper performance. *Newspaper researchjournal, 7*, 740-44, 792.

Maddux, D. (Ed.). (2003). *EditorandPublisher international yearbook: The encyclopedia of the newspaper industry: 2003* (83rd ed.). New York: Editor and Publisher.

Mandese, J. (2003, October 27). Report finds rich media 'ascendant,' even if ad impact is 'latent.' *MediaPost*. Retrieved January 14, 2004, from file:///Macintosh%20HD/Research-3/MediaDailyNews%2011-11-03%3Arichm

Mangalindan, M. (2004). Internet shopping…in person: New sites match buyers and sellers in same cities; good for cars, pets, houses. *The wall street journal*, pp. D1, D7.

Martin, H. J. (2003). Clustered newspapers operate more efficiently. *Newspaper research journal, 24,* 4, 6-21.

Maynard, N. H. (1996, June). Where is page one in cyberspace*The deeper news*. Emeryville, CA: Global Business Network.

Miller, B. (1998, February 21). Inevitable erosion of classified ads: The latest call to action in the online classifieds arena. *Editor and publisher, 131*, pp. 34-35.

Moore A. (1999, July/August). Small links, big chain: Community newspaper holdings and the powerof the cluster. *Columbia journalism review, 38*, pp. 13-14.

Morrill, D. (2004, June 15). IPIX emphasizes online ad tools. *Oakland tribune*. Retrieved June 28, 2004, from
http://www.oaklandtribune.com/Stories/0,1413,82~10834~2213769,00.html

Morton, J. (1997, March). A flurry of deals to create clusters*American journalism review, 19*, p. 52.

Morton, J. (1998, May). Small dailiesare dying in California. *American journalism review, 20*, p. 76.

Morton, J. (2004, February/March). Nouveau niche: Newspapers are once again turning to niche publishing. *American journalism review, 26*, p. 64

Morton, J. (2004, April/May). All pricesnegotiable: Newspapers used to have set-in-stone advertising prices. Now rate-card figures merely start the bargaining. *American journalism review, 26*, p. 68.

Newspaper Association of America (2003a). *Facts about newspapers2003.* Vienna VA: Author. Retrieved January 14, 2004, from
http://www.naa.org/info/facts03/highlights.html

Newspaper Association of America (2003b). *Facts about newspapers 2003*. Vienna VA: Author. Retrieved April 4, 2004, from
http://www.naa.org/info/facts03/20_facts2003.html

Newspaper Association of America (2004, April). Identifying the online audience: A new approach [summary]. Vienna VA: Author. Retrieved September 8, 2004, from http://www.naa.org/feds/onlineresource/onlineaudience.cfm

Newspaper National Network (2004a). NNN launches new top 25 and top 100 MRI codes. Retrieved August 28, 2004, from http://www.nnn-naa.com/

Newspaper National Network (2004b). We make newspapersnational . Retrieved August 28, 2004, from http://www.nnn-naa.com/about.htm

Newspaper National Network Posts Record Results. (1998, February 15). *The [SNPA] bulletin*, 3.

Niebauer jr., W. E.; S. Lacy; J. M. Bernstein; and T. Lau (1988). Central city marketstructure's impact on suburban newspaper circulation. *Journalism quarterly, 65*, 726-32.

O'Connell, V. (2001, November 16). Advertising: Web firms disagree on internetad hits. *The wall street journal*, p. B6.

OPA Intelligence Report (2003, November 24-December 5). *Social networking sites: The future of classified advertising?* Report from the Online Publishers Association.

Outing, S. (1998, March 21). Knight Ridder to launch nationalweb ad network: Rushing in to pluck business opportunities from the wreckage of new century network, KR teams with real media. *Editor and publisher, 131*, pp. 30-31.

Outing, S. (2002, October 9). Out with the old advertisers, in with the new. *Editor and publisher*. Retrieved April 25, 2004, from http://www.editorandpublisher.com/eandp/news/article_display.js...

Outing, S. (2003, April 30). Online business modelprogress report. *Editor and publisher*. Retrieved April 25, 2004, from: http://www.editorandpublisher.com/eandp/news/article_display.js... Also available as Online business model progress report. News sites bring in substantial revenues. Retrieved May 20, 2003, from: http://editorandpublisher.printthis.clickability.com/pt/cpt?action=cptandexpire=andurlID =6157570andfb=YandpartnerID=60

Outing, S. (2004, September 22). It's the end of classifieds as we know them. *Editor and publisher*. Retrieved October 29, 2004, from: http://www.mediainfo.com/eandp/columns/stopthepresses_display

Owen, B. M. (1975). Newspapers. *In economics of freedom of expression: Media structure and the First Amendment* (pp. 50-53). Cambridge, MA: Ballinger Publishing Co.

Picard, R. G. (1989). *Media economics: Concepts and issues* Newbury Park, CA: Sage.

Picard, R. G. (1993). Economics of the daily newspaper industry. In A. Alexander, J. Owers, and R. Carveth (Eds.). *Media economics: Theory and Practice* (pp. 181-203). Hillsdale, NJ: Erlbaum.

Picard, R. G. (2004). The economics of the daily newspaper industry. In A. Alexander; J. Owers; R. Carveth; C. A. Hollifield; and A. N. Greco (Eds.). *Media economics: Theory and practice*, 3rd Ed. (pp. 109-125). Mahwah, NJ: Erlbaum.

Picard, R. G., and J. H. Brody (1997). *The newspaper publishing industry.* Boston: Allyn and Bacon.

Potter, W. (1998, February). Blooming business: Newspaper clusters sow nondailies in the mix. *presstime, 20*, pp. 24-26.

Project for Excellence in Journalism, The (2004a). Newspapers: Audience. In The state of the news media 2004: *An annual report on American journalism*. Retrieved April 6, 2004, from http://www. stateofthemedia.org/narrative_newspapers_audience.as…

Project for Excellence in Journalism, The (2004b). Newspapers: Economics. In The state of the news media 2004: An annual report on American journalism. Retrieved April 6, 2004, from http://www. stateofthemedia.org/narrative_newspapers_economics…

Project for Excellence in Journalism, The (2004c). Overview: News investment. In The state of the news media 2004: *An annual report on American journalism*. Retrieved April 6, 2004, from: http://www. stateofthemedia.org/narrative_overview_newsinvestment.asp?media

Project for Excellence in Journalism, The (2004d). Overview. In The state of the news media 2004: *An annual report on American journalism.* Retrieved April 6, 2004, from http://www. stateofthemedia.org/narrative_overview_eight.asp?media=1

Project for Excellence in Journalism, The (2004e). Online: Economics. In The state of the news media 2004: *An annual report on American journalism.* Retrieved April 6, 2004, from http://www. stateofthemedia.org/narrative_online_economics.asp?cat=4andmedia=3

Project for Excellence in Journalism, The (2004f). Ethnic/alternative: Ethnic media overview. In The state of the news media 2004: *An annual report on American journalism.* Retrieved April 6, 2004, from:
http://www. stateofthemedia.org/narrative_ethnicalternative_ethnic.asp?media=9

Project for excellence in journalism, The (2004g). Overview: Economics. In The state of the news media 2004: *An annual report on American journalism.* Retrieved April 6, 2004, from http://www.stateofthemedia.org/narrative_overview_economics.asp?media=1

Rabasca, L. (2002). Plugged in: Newspaper online sites leverage audience data to improve content sell ad. *presstime*. Retrieved March 31, 2004, from:
http://www.naa.org/Presstime/PTArtPage.cfm?AID=6024

Ridder, P. A. (1998, February). Clustering: More than a fad. *presstime, 20*, p. 46.

Rodgers, Z. (2004, June 14). *A richer classified ad is born.* Retrieved June 28, 2004, from http://www.clickz.com/news/article.php/3367421

Rose, M. (2003, November 4). Wall Street Journal circulationrises 16% with web additions. *The_wall street journal*, p. B2.

Rose, M. (2002, March 18). Publishers take nips and tucks: Staffs, page sizes, coverage shrink because of slump; Will readership do same? *The wall street journal*, pp. B1, B6.

Rosenstiel, T., and Mitchell, A. (2004). The impact of investing in newsroom resources. *Newspaper research journal, 25*, 1, 84-97.

Rosse, J. N. (1980). The decline of direct newspaper competition. *Journal of communication, 30*, 65-71.

Rosse, J. N. (1975). *Economic limits of press responsibility.* Paper prepared for the Conference at the Center for the Study of Communications Policy Duke University.

Rosse, J. N., and J. Dertouzos (1978, December). Economic issues in mass communication*Proceedings of the symposium on media concentration, vol. 1* (pp. 40-192). Washington, DC: Bureau of Competition of the Federal Trade Commission.

Runett, R. (2004, February). Multimedia on the move. *presstime, 26*, p. 10.

Simpson, D. (2003, April 9). *More newspaper reporters may appear on TV.* Associated Press. Retrieved April 13, 2003, from
http://news.yahoo.com/news?tmpl=story2andcid=519andu=/ap/20030409/ap_on_re_us/asne_reporters_on_tv_4andprinter=1

Single-copy buyers: Attractive profile. (1996, August 10). *Editor and publisher, 129*, 7.

Solís, D. (2004, June 6). TV that speaks the language of young Latinos: English is the choice for new networkscourting prized group. *The Dallas morning news*, pp. 1D, 6D.

Stepp, C. S. (1999, September). Then and now. *American journalism review, 21*, pp. 60-75.

Stepp, H. E. (1996, August 16). Public transit use down, Study finds: Many commuting between suburbs rather than to cities, Census data show. *The Dallas morning news*, pp. A1, A14.

Stone, M. (2002, August 15). A time to compare numbers: Advertising measurement experts seek to standardize online traffic, print circulation and broadcast audiences (Electronic version]. *Online journalism review*. Retrieved September 9, 2004, from http://www.ojr.org/ojr/business/1029471539.php

Tillinghast, D. S. (1988). Limits of competition. In R. G. Picard, J. P. Winter, M. E. McCombs and S. Lacy (Eds.), *Press concentration and monopoly* (pp. 71-87). Norwood, N.J.: Ablex.

Tracking the online classified threat. (1996, December 7). *Editor and publisher, 129*, 23.

Web width guidelines approved. (2000, March/April). *TechNews, 6*. Retrieved August 21 2004 from www.naa.org/technews/tn000304/newsbriefs.html

Wrubel, C. (1998, April 20). Clustering: A technique for market domination. *Publishers' auxiliary*, p. 6.

In: Advances in Communications and Media Research, Vol. 3 ISBN 1-60021-188-7
Editor: Patula E. Grossmark, pp. 23-41

Chapter 2

THE TRUTH AND RECONCILIATION COMMISSION: MEDIA'S ROLE IN THE IMAGINING OF 'NEW' NATION

Charmane McEachern
Adelaide University, Adelaide, Australia

After 1994, the South African state had available to it few of the usual strategies through which societies construct themselves as nations. Apartheid had compromised language and culture as points of unity, and its association of race with nation had produced a country constituted through a complex of boundaries and cross-cutting ties, which defied any easy or inherited imagining of nation as 'the people'. It is on the basis of this that Thornton argues that the new democratic South Africa must be seen as postmodern (1996: 136-7):

> ...Apartheid as a formal system of laws and practices, failed because it constantly undermined the unity it sought through the implementation of difference. In its aftermath, the logic of difference remains, but it now lacks the political, philosophical and aesthetic image of unity of the (total) nation (1996: 142).

For Thornton, this postmodern condition precludes *any* imagining of nation, an assumption that is not only uncritical of the nation-produces-state formulation, but entirely underestimates the state itself and its need for nation as an expression of legitimacy and unity.

Rather than precluding 'the nation' as an imagined form of collectivity, Thornton's deeply entrenched logic of difference means that building the nation became a state project in pursuit of some version of multiculturalism, as the early slogans of 'unity in diversity' and 'rainbow nation' indicate. Difference itself required reimagining, no longer the source of otherness and hierarchy but now a fundamental building block of the nation. At the same time, through its attachment to race and role in the political power play of divide and rule, nation itself had become problematic under apartheid. The Truth and Reconciliation Commission (TRC) was thus a state project aimed at nation building which was to address all of these difficulties; the problematic nature of difference; the compromised concept of the nation and the difficulties of finding some new ground on which to construct unity through the constitution of 'the people'.

In one register, rather than seeking to establish some shared culture, the TRC was all about penetrating boundaries and divisions *through* the different experiences of different peoples, rendered as narrative. As a state-sponsored project, the aim of the TRC was reconciliation, reconciliation specifically with the potential to generate one, new, nation (Hamber 1998a: 5). In this sense, the TRC can be understood as an *enactment* of transition, legitimating both the state and the new government and enshrining a commitment to change (Hamber 1998a: 1), especially important where social and economic transformation was fraught with difficulty.

The second register is provided by what I have argued is the postcolonial context for the TRC performance of transition. Thornton's insistence on a postmodern though not postcolonial identification for South Africa (contra Pechey 1994) is problematic because it ignores the ways in which non-white South Africans were precisely disenfranchised, organised and subjugated to activate a form of internal colonialism, itself a product of colonialism and thus already postcolonial (1993: 295).[1] The apartheid concept of nation was problematic in this context because it too was a structural element of this internal colonialism, rendering regions which had once been integral parts of the country simultaneously 'foreign' and dependent, while at the same time supplying necessary labour. In this second register, the TRC was an excellent example of the postcolonial requirement to predicate reconciliation on a notion of 'truth'; discovering the truth about the past, interpreted as giving name and voice to those experiences which had not been heard or which had been denied, through the lies and cover-ups which were so essential to the rule of the apartheid state (Tutu 1999: 32, 33; Hamber 1995: 1). Together, these two registers provide a ground for a 'new' South Africa; penetrating divisions and revealing truth constituted the TRC as a necessary moment between past and future, an extended present in which to begin a new imagining of people and nation.

From the perspective of the post-apartheid state, through the concept of 'truth recovery' the TRC is an example of the quite intentional *development* of nationalist discourse emerging "in dialogue with colonialism", of which Chatterjee speaks (cited in Nagengast 1994: 119). Its broadest strategy was the construction of a memory culture of which Huyssen says:

> ...today's critical memory cultures, with their emphases on human rights, on minority and gender issues, and on reassessing various national and international pasts, go a long way to provide a welcome impetus for the writing of history in a new key and thus for guaranteeing a future of memory (2000: 36).

At *this* level, it did not matter whether ultimately the result was only partial, containing further silences (Bozzoli 1998; van der Merwe and Kgalema 1998: 1). This is why one television producer was able to say that *process* was as important as specific details of who did what to whom (Du Preez on *Late Night Live*, 15.9.99). In its sustained and focused talking about the past, the TRC offered up the nation as its object of contemplation, or more accurately brought the embryonic nation into being, since before 1994 people inhabited different nations. The TRC constituted itself as a ground for imagining *the nation*, and the specific symbolic form of imagining was the telling of stories to make connections between people once understood to inhabit different spaces, different histories.

---

[1] Pechey also suggests that the term neo-colonial could usefully describe apartheid rule here (1994: 151-57).

In the postcolonial context of South Africa in the 1990s, the TRC went further than just confronting the past in order to heal or provide a moral basis for political action, two of the most often cited aims (Liebenberg and Zegeye 1998: 546). Its version of making "history in a new key" out of a memory culture (Huyssen 2000: 26, 36), was to *harness* disparity and produce unity in the form of a new official history; a narrative (White 1980) of the country in which all people have a place (whether they agree with it or not). Such a narrative of the past was itself a beginning in imagining a new nation. Through this new official history, the TRC provided one avenue of legitimation, albeit contested, for the postcolonial state. This new history challenged the past regime, not just in content but also in terms of form. At the same time the TRC itself *stood for* the rupture that Werbner (1998, see Chapter 1) notes new states in Africa have often used to symbolise new nation. The TRC was directly implicated in the claim that this South Africa is new.

If all that has been said so far is correct, then this vehicle of unification and legitimation must have a very public form; nation must emerge from what Huyssen calls a public memory culture (2000), a highly visible process. Hayner (cited in Liebenberg and Zegeye 1998: 550) asserts that the South African Truth Commission is "the first example of a process officially opened to encourage public debate and input". This chapter argues that in South Africa this need for a public face, for high visibility, generated a fundamental role for the media in producing a public TRC with the potential to contribute to the construction and shaping of a nation. Itself a public form, the TRC was amplified and shaped through media of all kinds (cf Werbner 1998: 15). As such, the media TRC in South Africa was far more than an example of modern historical life taking place in the media. Rather its media form emerged from and was an essential ingredient of the process that the TRC offered to South African nation building and the state. The TRC that most South Africans knew was fully mediated, accessed only through the press, television and radio, all of which covered it extensively. It was the media version that they engaged with, criticised, approved of and used in their own new narratives and performances of place and identity (see Chapters 4, 5 and 6). What emerged was a media TRC which is also performance; constructed for an audience and shaped by and dependent upon media agendas, genres, discourses, conventions and forms of communication, the politics and economics of media and the professional ideologies carried and practised by those bringing it to the people. Thus the new history itself was mediated or media produced history.[2] Its media form made it possible for the TRC publicly to enact transition; producing and promulgating a new official history - even though many people refused to involve themselves directly in the process itself[3]

This chapter introduces the TRC as it was mediated and performed in media of all kinds, from print and radio to television and the Internet. It argues that in many ways the TRC can be understood as a media event, though its long running nature and the involvement of all media necessitate analytical modification in how one is to understand media event. The chapter begins with the establishment of the TRC, in particular exploring the role of media in

---

2 For a discussion of this aspect of media see Dayan and Katz 1992; Street 1997; Stevenson 1995; Thompson 1995.

3 See Brink 1999; Krog 1998; Theissen and Hamber 1998; Tutu 1999; Hamber et al 1998: 3; Hamber 1996b. In a long article in the *Mail and Guardian* (20.3.97), American writer David Goodman commented on this refusal, contrasting it to the TRC's success in revealing "the truth behind many major massacres and high-profile political assassinations". "And yet most whites have turned a blind eye to the testimonials of injustice, just as they did in the past", he said. Nancy Scheper-Hughes (1998: 11) also argues that "a great many white people in South Africa still fail to get the point behind the TRC". See also Don Mattera's comments in Jones (1999).

its processes in order to set up the next chapter's discussion of the content of the TRC media event. The discussion seeks to demonstrate the way in which the imagining of the 'new nation' set in train in the TRC was fundamentally shaped by media's own agendas and forms. The argument is based largely on media texts, newspapers, television and radio, but also draws on comments made by journalists both in media and in personal exchanges and on discussions with many people in South Africa not directly involved in the TRC at all.

THE TRUTH AND RECONCILIATION COMMISSION

Reconciliation is very important for the ANC led government in South Africa. Interpreted in various ways, it has become one cornerstone of the new state and society, a language of, and strategy for, the unity of nation. The TRC is the most comprehensive and visible manifestation of this commitment to the role of reconciliation in transformation. Created as an act of Parliament (The Promotion of National Unity and Reconciliation Act) in 1995, the TRC was conceptualised in terms of "a truth recovery process that is aimed at reconciling with the past" (Hamber 1995: 1). Initially at least, truth was given a fundamental place in the process of reconciliation (a formulation which many commentators (Hamber 1997a and b, 1998b) and participants (Tutu 1999) note ultimately proved problematic and reconciling the nation was at least as important as individuals.[4]

The TRC was a public event, in the sense meant by Handelman (1990: 3), who describes such occasions as ones that "people undertake in concert to make more, or less, of themselves, than they usually do". Bozzoli (1998) emphasises the ritualised elements of the TRC as a public event, carefully crafted to heal individuals but also to make the transition between the old and the new in society. As a public, ritualised event, the TRC was also a *political process*, a fundamental element of the negotiated settlement which ended apartheid.[5] At the 1992 Codesa meetings, it was agreed to create a truth commission as a way of avoiding stalemate and the continuation of violence and civil war, and the fact of its creation was contained in a postamble to the interim constitution of 1994.[6] In many ways the TRC is symptomatic of that settlement and problems such as the incompatibility between justice and political expediency at least partially derive from these compromise conditions of its creation.[7]

The negotiated end to apartheid ruled out a purge of, or special state trials for, apartheid officials, some of whom became members of the Government of National Unity (1994-6). Yet at the same time, other parties found totally unacceptable the National Party's attempts, while still in power, to grant blanket amnesties to those involved in the regime's rule. In

---

[4] Hamber (Manager of the Transition and Reconciliation Unit at the Centre for the Study of Violence and Reconciliation) for one, argues that here the TRC has been more successful at the national level than the individual, and that far more work has needed to be done to help individuals to reconcile (see also Hamber et al 1998; van der Merwe 1998)

[5] See Tutu 1996: 41; Asmal et al 1996; Hamber 1998a.

[6] It should be noted that the postamble specifies the creation of an amnesty process. It was the work of politicians and civil society to devised the gross human rights violations hearings to run in conjunction with this amnesty process.

[7] During many debates about the problems of the TRC, participants like Alex Boraine (Deputy Chairperson of the TRC) returned again and again to the fact of the negotiated settlement as the primary context for the constitution of the TRC.

choosing the TRC option, inspiration came from other truth commissions, especially the fifteen held since the 1970s. Personnel from some of these commissions, those in Chile, Argentina, El Salvador, Rwanda and Bosnia, all had input into the process[8] out of which the South African truth commission emerged. At the end of this process, options such as state run or internationally conducted commissions were rejected in favour of a separate commission set up by the state but independent from it. Suspicion about the reality of this independence fuelled criticism and non-cooperation from parties like the New National Party (NNP) and the Inkarta Freedom Party (IFP) and a range of, particularly white South Africans, throughout. Indeed, from the time that it left Government in May 1996, the NNP's polarising rhetoric and accusations of witch-hunts,[9] coupled with a refusal to co-operate wholeheartedly, became strong strands in both the TRC and its media coverage.

Although the appointment of the 17 person Commission was the job of the President, commissioners were largely chosen from a list of nominees and potential commissioners were subjected to a public examination process (see Tutu 1999: Chapter 5). Meetings were held all over the country for people to have their say and NGOs, like the Centre for the Study of Violence and Reconciliation (CSVR) and civil society, not only facilitated debate but also scrutinised the TRC throughout (Hamber, Mofokeng and Simpson 1997). Media became a primary arena for debate with editorials and letters pages providing criticism and commentary on a daily basis. The process of setting up the TRC quite deliberately established principles of transparency, openness and visibility at the outset. Although it reinforced the postcolonial nature of the process, inevitably this characterisation was only partially true.[10] Hamber and others (1997) argue that in fact, prior consultation with civil society bodies and individuals and an extensive civic education program would have greatly reduced suspicion about the process (cf van Zyl 1995).

In many ways it was the role of media which both provided the sense (and promise) of openness and inclusivity as well as their shortcomings. Not just the hearings but the education process itself was mediated. In addition to providing a vehicle for transmitting news about the state of progress of the TRC, media (especially radio) were used to educate people about the TRC. Of this education campaign Tutu says:

> It was important to reach out to as many South Africans as possible. Our aim in fact was to reach out to all South Africans; to engage them in the work of their Commission and to ensure that none would lose out by default, that all would realise that they had the opportunity to tell

---

[8] See Boraine and Levy 1995; Boraine, Levy and Scheffer 1997; Liebenberg and Zegeye 1998: 550; Edelstein 1999; G. Simpson 1994a and b; also "The Rights of the Wronged", last in the documentary series *If Truth Be Told* (SABC3, 3.6.96). The Justice in Transition Project, of which Alex Boraine (later Deputy Chairman of the Truth and Reconciliation Commission) was the executive director, ran conferences at which the kind of commission required by South Africa was extensively debated.

[9] The accusation of "witch-hunts" was not limited to the NNP. In a study of whites' attitudes to the TRC for the CSVR, Theissen and Hamber report that, "A significant number of white South Africans surveyed (46%) believed that "the TRC was an ANC-inspired witch-hunt to discredit its enemies, although the same percentage disagreed and 8% had no opinion" (1998: 3).

[10] Thus we find in an article on the work of the Commission's staff at the end of the Commission ("On a wing and a prayer: How the staff survived", *Mail and Guardian* 30.10.98), the staff talk about the secrecy and the discretionary decisions about who they would actually talk to and from whom they took statements. They characterised this as an inevitable outcome of the process in which the 1995 Promotion of the National Unity and Reconciliation Act of South Africa which generated the TRC had actually to be applied "on the ground" in their work.

> their stories and might be eligible for reparations. Our advertising campaign focused especially on the radio, which was more effective in reaching illiterate people. We put out pamphlets and posters with a distinctive logo and with slogans such as, 'The Truth Hurts, But Silence Kills' (1999: 81).

Media campaigns such as these meant that the TRC could be represented in terms of openness and inclusivity, providing postcolonial narrative strands for the new history which themselves depended on media exposure and representation.[11] The public (including media) forums and information, operating alongside a notion of accountability, not only suggested change in South Africa, but also separated South Africa's TRC from other truth commissions, many of which the powerful had used to protect themselves and their interests. Ordinary people would be able to make *this* Commission work for them, which is why the very *character* of the process was seen as transformative. This was both a political and a moral proposition, emerging out of the liberation ideology of participatory democracy and the discourse of healing which Scheper-Hughes argues "has emerged as the master narrative of the late 20th century, as individuals and entire nations struggle to overcome the legacies of suffering..." (1998: 9). Archbishop Tutu, who became the Chairman of the TRC, was particularly adept at pulling all of these strands together, becoming a major media figure in this public process (see for example Tutu 1996).

There were three committees in the TRC; the Human Rights Violations Committee, the Amnesty Committee and the Reparations and Rehabilitations Committee. The first two held public hearings, convened in the regions in which the events had taken place. The most visible in the media, the Human Rights Violations Committee documented apartheid's crimes. Submissions were received in writing from victims and a percentage of the around 22,000 received was selected for the actual hearings which began in East London on April 15, 1996. Only 2,000 cases were actually heard, so that a relatively small percentage of the acts of violence and abuse became the public account of those years.[12] Hearings were a mix of well known and obscure local cases (cf Bozzoli 1998: 168).

The Human Rights Violations committee aimed to confront the silences and denials which had constituted and enabled the apartheid regime's practice of power and domination. The hearings thus not only displaced colonial relations and practices in exposing apartheid's oppression but sought to highlight and give voice to experiences silenced in official accounts of South African society under apartheid. They were predicated on the need for the victims to speak from within subjectivities not readily constituted in dominant colonial discourse, allowing people to tell their stories in their own languages and styles, with translators making these widely accessible. For some of the relatives of even the most well known of the violations, this was the first time their story had been told in their own terms. One woman, one of the mothers of the so called Guguletu Seven, made this clear when, in Xhosa, she explained that the family had not even understood what was happening at the inquest into the death of her son at the hands of security police, because the entire proceedings had been

---

[11] Public education about the TRC was also facilitated by NGOs like the Centre for the Study of Violence and Reconciliation (CSVR). The CSVR ran workshops on the TRC during the whole process, particularly with a view to giving people information about the TRC but later seeking information on the effect of the TRC in their local region (Hamber et al 1998).

conducted in Afrikaans without translation (SABC 3 News, 23.4.96). Part of people's feelings of oppression under apartheid was that they never felt that they could understand and negotiate bureaucratic procedures (Bozzoli 1998) and the very fact of telling their own stories in a forum which gave them precedence and respect was both self-affirming and a demonstration of change in society.

This was not the first time that the stories of the oppressed had been heard directly. For example, Thale (1995) explores the autobiographical accounts of four black workers in the 1970s and 1980s but notes also the critical silence with which these texts were received. From a different perspective Coullie (1997) notes that in the early 1980s radical academics had recorded and published stories of rural workers, women, domestic labourers and so on, in an attempt to provide a voice for the powerless under apartheid. The TRC was different in several ways; in its aims of reconciliation and healing and in the encompassing objective of setting up the nation for contemplation and healing to which these acts of narrating the past were directed. Both of these aims produced a third and primary difference from past attempts to voice the narratives of the oppressed; the *public* nature of the TRC's voicing[13] and its sanctioning by those in power, particularly the state.

The media were profoundly implicated in these differences; they not only extended and legitimated the TRC as a public forum but gave it a popular and readily available form in television, radio and newspaper. If one takes the TRC as a site for imagining the new nation and uses Bhabha's distinction between the pedagogic and performative registers of nation, then the TRC as the statement of nation in a pedagogic register quite deliberately and consciously *harnessed* the performative, in the form of the narratives of the people (cf Bhabha 1990: 297- 308; cf Werbner 1998: 8). At one level this could be seen as an appropriation of the performative to signal transformation; nation enshrining diversity, inclusion, democracy and participation. Certainly the ritual structure and features of the TRC hearings sought to encompass the diversity of the narratives in the frames of truth and reconciliation. At the same time, the central role of media in elaborating and amplifying the inclusivity of the TRC also gave them a pedagogic function, allying them to the aims of the state. Their performances too signalled transformation as we shall see.

The amnesty hearings began in May 1996. Contentious was the fact that, within the frame of restorative rather than retributive justice, confessing perpetrators did not have to apologise, only to make full disclosure and prove political, not personal motives (Marx 1998: 218; Miller 1997; Dissel 1997). In addition there were a number of institutional hearings, with bodies like the media, business and the health sector testifying to their role in supporting and legitimating apartheid's abuses. Amnesty hearings were permanently framed by the perceived inadequacies of negotiated settlement, with many people never able to accept confessed murderers walking free immediately after the granting of amnesty, particularly while their victims had still to wait for reparations. Further, the fact that the most of those claiming amnesty were already in jail cast initial doubt on the ability of these hearings to generate the

[12] For example, if we take the hearings in just one place, the township of Alexandra in Gauteng, Bozzoli tells us that only 22 of the enormous numbers of people who were apartheid's victims in this most troubled of townships, were invited to testify (1998: 167).

[13] Hamber (1996b) notes that voicing was even important for white victims of violence. He notes one of the Church Street bombing victims saying, "the new government is significant to me because now I can sit here and tell my story", and comments, "If she feels this, and she probably had more space to speak out than the majority of those victimised in the past, the initial hearings must have been successful in giving a voice to the victimised."

new information about apartheid society which would be necessary if a new history was to be created. The hearings largely featured 'foot soldiers' while major leaders like de Klerk and P.W. Botha, and Winnie Madizikela-Mandela refused to apply, generating a credibility problem for the Commission throughout.[14] And finally, there was contestation around the cut off date for amnesty. Though the starting point was always 1960, the year of the Sharpeville massacre, the cut off point was eventually extended to May 1994 to include some of the, particularly right-wing, activity to destabilise the elections.[15] Contestation here was in part also generated by the real fear that the killers of Chris Hani would thereby be able to seek and get amnesty. They did apply, but it was not granted. The frame for the hearings very much functioned to separate apartheid out from the pre-apartheid systems which provided the forerunners to the institutions and processes of apartheid.[16]

Reparations are the focus of the third committee and here, even after the final TRC report has been published, much work is still to be done. Problems are legion, from the lack of funds to the enormous difficulties of deciding just how a society can possibly recompense the loss of life or the injury to body and mind which permanently incapacitates a human being. Should reparations be monetary or symbolic, at the level of individuals or communities? Just who should qualify?[17] The Reparation and Rehabilitation Committee had no public hearings, operating rather as a committee of review to make recommendations to the President. The TRC's approach to reparations changed somewhat over the period of hearings, from an initial disinclination to consider individual reparations of any significance to recommendations of substantial payments to victims over a six year period with emergency payments in the interim (see Tutu 1999).

If the media had a central role in informing people about the TRC, it also became a forum for the inevitable contestation which emerged around the process, before, during and after the hearings. In counterpoint to their pedagogic register, though also affirming this, there is always space in media representation for alternative responses, though this will always be contained within particular limits. In South Africa, in the face of poor consultation at the level of civil society which concerns researchers in the CSVR (see above), the media provided one clear arena in which (and to which) people could publicly respond.[18] Here, as healing and speaking became major, enduring TRC themes, the narrative strands of guilt and accountability also emerged and persisted, most notably through the ongoing objections of

---

[14] Even after all of the hearings, this remained a problem, as the *Mail and Guardian* editorial (13.11.98) points out. See also du Preez in Jones (1999). Of course, as many analysts noted, it raises the very real problem of what should be done next.

[15] In his account of the TRC, Archbishop Tutu describes the cut-off date, which was the 10th May, the day of President Mandela's inauguration, as appropriate because of its symbolic nature as the day of the beginning of the new South Africa. "If any event might be said to mark the advent of the new dispensation, it was that day which demonstrated an irrevocable break with the past of conflict, alienation and domination of the many by the few," he said (1999: 78). In this conception, the TRC is given meaning by becoming a central element of the narrative of a new beginning in South Africa, which encodes the nation in terms of time (cf Handelman 1990: 223- 33). This temporal dimension was certainly significant in the pedagogic register of national imagining which the TRC also represents.

[16] For discussion of this aspect of apartheid see Mamdami 1996; Mabin 1986; Norden 1994.

[17] The CSVR has produced a number of extremely useful papers and articles on this very difficult area of reparations (see for example Hamber 1998b; Hamber and Rasmussen 1999).

[18] In this register one could see the media as an arena for the clash of elite and popular resolutions of memory crisis which Werbner explores in a different context in Zimbabwe (1998: 97). Certainly these exchanges and

those like the families of murdered activists Steve Biko and Griffiths Mxenge, for whom the healing relationship between truth and reconciliation demanded justice in the law courts.[19] In a different vein, South African author Phyllis Jordan attacked the TRC as "an insult to black people…aimed at white-washing the sins of whites" (*Weekend Argus*, 15/16.6.96).[20] There was some dissatisfaction voiced among those, like the infamous Dirk Coetzee, who had served the apartheid regime too.[21] And Robert McBride, the MK cadre responsible for the Magoos Bar bombing in Durban, became the centre of a different kind of attack on amnesty, when the two sisters of one of the victims bitterly attacked his freedom and Government position on the grounds that this was benefitting from human rights violation.[22] Thus the actual meaning of reconciliation, justice (retributive or restorative) and the problematic relationship between them emerged as issues early on, not just in the conferences setting up the legislation, but also in the public media debates which sought to interpret, interrogate and elaborate the dominant frames established for the TRC. Another strand which also emerged, often publicly through state agents of apartheid (for example in the police submission to the conference held by the Justice in Transition Project in 1995), but taken up by both victims and perpetrators of apartheid, interrogated reconciliation differently by asking whether truth and memory or forgetting best served the interests of reconciliation.[23] This strand also remained in public discourse once proceedings began, providing a constant need for pro-Commission people, both inside and outside of the TRC, to engage with it.[24]

The presence of the media in these comments on the TRC process alerts us to the fact that, for the majority of the population, the whole process, so much bound up in the building of nation, only existed through media. Extensive media interest and coverage brought this ritual event, which was also a political process, into people's homes and thus made it part of their routine worlds. Media provided an exposure to, and form of participation in, the TRC

interventions in media demonstrate very real contestation over memory and the role of memory in postcolonial societies.

19 See also "Last gasp of a turbulent era", *Mail and Guardian* 24.12.98 and "Commission creates fresh interest in past", *Mail and Guardian* 3.1.97. On this issue, these radical refusers of reconciliation in this form had something in common with (though not in terms of interests) the NNP which officially and consistently refused to accept the legitimacy of the truth commission process, also insisting that any 'transgressions' be seen as law-breaking and be dealt with accordingly.

20 Jordan's comments were doubly media worthy because she is the mother of then Minister of Environment and Tourism, Pallo Jordan.

21 See for example, "Amnesty fails South Africa's unforgiven", *Mail and Guardian* 30.8.96.

22 See "Has ex-MK man McBride become a scapegoat", (*New Nation*, 24.5.96: 11); "Commission puts down roots of reconciliation", (*Mail and Guardian* 10.5.96). McBride subsequently applied for amnesty although he did not need to, having served a prison sentence for his crime. Interestingly, his amnesty hearing revealed more bombing attacks which had previously been unknown in connection to him, so again the TRC process could be said to be successful in finding out 'the truth'.

23 See Scheper-Hughes 1998; cf Parry 1995/6; Hamber 1996a, 1998a, then Werbner 1998: 96 for the same kinds of debates in the context of Zimbabwe.

24 See for example, Krog in the *Arena* documentary, "Stories my Country has Taught Me"; "Rights of the Wronged". Both in this documentary and in the 1994 conference held by the Justice in Transition project in Cape Town, Krog defended the TRC and attacked any notion of forgetting the past from the position of learning from Afrikaner history and identity. At the conference she in part attributed apartheid to the failure to hold some form of truth commission after the Boer War:

British abuses against women and children were recorded by Afrikaners but never officially acknowledged or condemned by the British. Thus the tales of the war did not become part of an ethos relating to how people should behave towards one another. Rather, they became a folklore supporting the notion of Afrikaners as a threatened group and a belief that any behaviour, however outrageous, was acceptable if it fostered their survival (Boraine and Levy 1995: 115).

which it could never have had through direct contact, even though the hearings moved around the country. Hamber (1997a: 1) argues that through media "in the urban areas almost the entire population must have been exposed to the work of the TRC and consequently the stories of victims; this is less certain in the rural areas". In many ways here, one could argue with Becker (1995: 640), that it is media which actually turns a process like the TRC *into* a public event, since without media it would be far more limited in scope. As Becker states,

> Although media do not offer a valid substitute for the experience of participation in the event, they do extend the possibility of reflecting on the significance of the cultural performance to those who were not present (1995: 642).

It was clearly this possibility which in part underpinned the TRC education through media process which Tutu described earlier and we shall see later that it was also this broadening of the base of reflexivity to which journalists responded in their coverage of the TRC. At the same time, media's operation as a primary source for information about the TRC also inevitably influenced the public perception of the process, since some parts were more accessible or newsworthy than others, becoming significant dimensions of the *media TRC* at the expense of others. In this media staging of the TRC, we have the performative being introduced into the pedagogic dimension of nation at a different representational level, with important consequences for the imagining of nation. It is therefore necessary to consider the actual work of the media in producing a public TRC event. I want to do this by way of a discussion of the role of the journalists who had the task of producing and constructing the TRC as a public event into its media (or mediated) form.

THE MEDIA TRUTH COMMISSION: JOURNALISTS AS CULTURAL BROKERS

In Chapter 1, I argued that media of all kinds had been deeply implicated in apartheid's rule; they often colluded in and legitimated the system, demonised, ignored or silenced the very people now invited to speak at the TRC proceedings and they obscured, misrepresented or simply denied precisely those offences which formed its subject matter. The TRC was thus a watershed for many media institutions and personnel. The media hearings in September 1997 demonstrated that, despite differences in degrees of support and debates over the relative culpability of journalists, editors and owners, media were inextricably part of the apartheid system (Burnett 2000; Hamber 1997a; Brynard 1997). Editors and journalists working directly to support apartheid and the National Party was one thing, but the hearings also heard how even liberal media workers at times uncritically accepted government versions of events, censored their accounts, demonised organisations such as the ANC and UDF as 'terrorists' and communists and opposed anti-apartheid activities like sanctions and consumer boycotts. Further, black media workers were routinely and systematically discriminated against.

Given that the TRC had in part a responsibility to establish new moral principles on which to ground not just the nation but also society's institutions (Simpson 1994a:2; van Zyl and Simpson 2000), it also raised the need for media, in particular journalism, to transform themselves and provided the opportunity to do so.[25] So, in 1996, reporting the TRC became part of a new beginning for journalists, as responsible, independent and critical news sources (Krog 1998: 19,20), who could also overcome the "credibility gap" which Burnett (2000: 14) argues was now a major audience perception. This gave the TRC a doubly political form; it was a political process, but one that also offered some news media at least, the opportunity of making reparation. *Reporting* the TRC *also* became a performance of change.

In reporting the TRC, media personnel operated as cultural brokers, working at the interface between the TRC proceedings and personnel and the public, now constructed as *audience*. Bourdieu (1986: 359) has drawn our attention to this crucial role in "presentation and representation" of media personnel, arguing that their job in society is the very important production of symbolic goods (see also Thompson 1995). They select elements of culture, interpret them, frame them and make sense of them (and through them), for the audience. That is, they 'broker' culture, and in dealing in it, they also *reveal* it (in highly mediated ways) to us (see also Peace 1998). Through a range of strategies, they show their audiences how to make sense of events and people, what (and who) is important about them and what they mean (Bourdieu 1996). They set agendas and produce knowledge by categorising events and people. They are in the highly privileged position of having their representations and interpretations heard but also become part of the very constitution of the world. In reporting the TRC, journalists and other media personnel brokered the public event in all of these ways.

It was as representatives of the people that media approached their brokerage role. Here, media positioned themself through a usual professional practice. As Kumar (1977)[26] argues, assuming the role of representative of the public with a duty to expose and illuminate events on behalf of the public, is a common authorising strategy for journalists and current affairs presenters. From this position, they speak *for* the public, which is exactly what journalists did in the media TRC when they asked questions about reconciliation, truth and justice and so on. It is a unifying strategy in a populist vein, regularly deploying the encompassing language of 'we' and 'us' that influences the representation of the event. For example, journalists referred to "our past", a persuasive strategy which, in the context of South Africa and the TRC, also obscured the very differences emerging in the hearings, establishing an inclusive us whom *journalists* represent. What is interesting in media's representation of the TRC is the way in which this professional journalistic strategy also coincides so profoundly with the discourse of nationalism, particularly its need to construct and invoke a unified version of 'the people'. It actually constructs and naturalises a we on which nationalism can be built. In journalists' performance, it speaks a nation where actual certainty of nation is not at all sure (cf Bhabha 1990), which is exactly the point of the TRC. Of course, like all media attempts at achieving preferred meaning (Hall 1980), such strategies were not always successful and there were people who refused the attempt at persuasion.[27]

---

[25] The Commission offered this possibility also through the 1997 media hearings which were no less contested than those of other institutions under apartheid, such as medicine, business and the law (Burnett 2000). In just one example, *Nasionale Pers* decided that they had nothing to say to the TRC, a decision which was both supported and opposed by its journalists, 127 of whom submitted statements to the Commission apologising for their role in supporting apartheid (see Brynard 1997; contra van Staden 1997).

[26] See also Thompson 1995: Chapter 5; Street 1997; Morse 1993, 1998.

[27] This is something that even that great optimist Bishop Tutu came to accept. In one lengthy interview with the *Mail and Guardian* he says as much; But, he reflects, perhaps it is time to be more honest in recognising differences in perspective. Instead the country is sliding away from confrontation, toward cosy positions, and when they get there they are discovering that they are in different worlds. Blacks resent the failure of whites to

Despite ongoing interrogation, through their actions as brokers, media personnel seem largely to have accepted both the nation as the object of contemplation for the TRC as a nation building process, and the legitimacy of this as a state project. In taking this brokering position, journalists then took up the pedagogic imagining of nation embedded in the TRC design as a public event (Handelman 1990: 17). Here, they also seem to have accepted the early orthodoxy of 'truth' before 'reconciliation', which itself had a construction of nation built into it. I will argue in the next chapter that this exactly suited the conventions of news which were fundamental in the media TRC. For now, it is useful to note how the deliberate choice among some journalists to use the name "Truth Commission" (Krog 1998: 45), is one example of such acceptance, embedding the official arguments that South Africa cannot face the future without knowing the truth about its past, and its primary concern to generate a new narrative about the past. The journalists who dominated the Commission's coverage seem certainly to have accepted this orthodoxy. Max du Preez, executive producer of the excellent weekly *Truth Commission Special Report* on SABC2 made this clear on many occasions. For example:

> We were driven by our commitment to the story. Most of the team were young and inexperienced in television journalism. But they felt they owed it to South Africa to tell the story.[28] This is where our roles as journalists were blurred into our roles as citizens. We felt it was a massive privilege to record our history for our children and grandchildren (Thloloe 1998: 2).

and later,

> We were making the truth a South African burden and taking it into people's homes (ibid: 3).

As cultural brokers, journalists can be understood to operate within a political field (Bourdieu 1991) which has increasingly produced a *mediated politics* in the sense meant by Altheide and Snow (1991) as the fields of politics and journalism interact. Altheide and Snow are describing the emergence of a politics which has responded to what Bourdieu (1996: 57) calls media *heteronomy* - the penetration of the political field by that of media. Like many political processes, the TRC had its most visible and accessible form in the media, to the extent that aspects of the Commission had themselves responded to media needs. Press releases informed and generated news stories, personnel used their own press officers and to an extent the political process took up the cult of celebrity, with figures like Archbishop Tutu coming to almost stand for it. At the same time, the notion of mediated politics, though valuable, cannot really encapsulate the full dynamic of the relationship between the TRC and the media, nor the interpenetration of the two, such that a new kind of mediated historical event was being created in South Africa. We have established that the TRC needed media and that media became co-producers of it. At the same time, the TRC itself is an event of a different order from the elections and political manouevrings which Altheide and Snow use to

---

realise how lucky they are. Whites resent the burden of guilt they carry for having benefitted from the system ("Fisher of men, seeker of truth", *Mail and Guardian* 19.12.97).

28 Note media's immersion in the language of narrative and narration as a way of characterising the TRC (and the new South Africa) here. Media were very active in helping to produce the climate in which people in all walks of life felt the need to 'tell their stories', which was noted in the introduction to this book.

elaborate their category of media politics. It is open-ended rather than closed, unpredictable and far more deeply formative than these examples it is an integral element of transformation rather than a restatement of a given political situation. Here, the literature on media event would seem to be a better starting point, while the TRC as a case in turn allows us to interrogate this literature.

THE TRUTH AND RECONCILIATION COMMISSION AS MEDIA EVENT

In the last decade, work on media events has emerged out of anthropology, from work on festivals, ritual and ceremony extended into public events. Two of the most influential have been Dayan and Katz, whose book, *Media Events: The Live Broadcasting of History* (1992) has influenced a whole range of studies.[29] Dayan and Katz argue that today, historical and political public ceremonies are actually media events; that is, staged in media form and therefore deriving much of their ritual and symbolic content from media genres and conventions. For them the media event is "a new narrative genre that employs the unique potential of the electronic media to command attention universally and simultaneously in order to tell a primordial story about current affairs" (1992: 1). Their focus is on television. Such an event is a ceremony within the political field that also has the ability to transform audiences into the public, whose presence in front of the television set becomes an essential element of the event itself.

Dayan and Katz use examples as widely ranging as the funeral of John F. Kennedy, Anwar el-Sadat's journey to Jerusalem in the name of peace, royal weddings, the Olympics and the Watergate hearings, generating a global typology of media events, conflicts, coronations and conquests, which corresponds to the forms of authority which are legitimated and enacted in each one. For these authors, as a class media events differ from news genres by virtue of various common attributes. They are live ceremonial occasions, characterised by interruptions of routine (regular broadcasting including advertising is suspended), the organisers of the event are public bodies (often state authorities) and they are "*preplanned*, announced and advertised in advance" (1992: 7). Media events are in a sense co-produced, by organisers and broadcasters and the latter typically suspend the critical dimension of cultural brokering and approach the event with awe and respect. Television, say Dayan and Katz, "upholds the definition of the event by its organisers, explains the meaning of the symbols of the occasion, only rarely intervenes with analysis and almost never with criticism" (1992: 8). It is this *tone* which also invites the audience into the necessary role of the public for the event, a public which is enormously expanded by the ceremony's access through television. Media events' audiences are huge as viewers suspend normal life and gather in front of the television to participate in them (1992: 9).

Many of these features, live broadcast, suspension of critical function and of everyday life, for Dayan and Katz preclude the news, even great news events like a political assassination, from being a media event. "Great news events speak of accidents, of disruption; great ceremonial events celebrate order and its restoration," they argue (1992: 9). It is this narrowness in defining media events which has been the subject of debate. The reservation of

---

[29] See for example Jacobs 1992; Becker 1995; cf Scannell 1995.

such an important term for a relatively small number of ceremonial public events is highly problematic when today, in many societies, a situation of media heteronomy in the political field is the norm. Today most political processes are mediated, performed if not wholly staged in a range of ways for media of all kinds (Scannell 1995). This *range* of political processes is actually embedded in Dayan and Katz' examples as Scannell notes, causing him to criticise their typology for failing to distinguish between "mediatized public ceremonial occasions and mediatized public processes" (1995: 5). It is here that Dayan and Katz fundamentally underestimate the role of media, especially the news and current affairs, in political process today. Their examples of events in the news, such as earthquakes or disasters, exacerbate this tendency. Yet, contemporary political processes and events, as historical processes or processes making history, are basic news material, and these are very much staged and performed with news media in mind, as Altheide and Snow demonstrate very clearly (1991).

If one takes the TRC as a case in point in this debate, the first week of the hearings could indeed be categorised as a media event in Dayan and Katz's terms. The hearings were broadcast live on television as well as radio, extensively advertised beforehand and by all accounts, provided engaging viewing, with tens of thousands of people stepping out of the routines of everyday and into the TRC through their televisions and radios. Yet, the proceedings thereafter were only broadcast live on radio and the TRC became more "mediatised political process" (Scannell 1995), still covered extensively in news, current affairs, newspapers, magazines, documentaries, studio discussions and forums (radio and television), commentary programs of all kinds and so on. To argue that it ceased to be a media event at this stage, ignores several factors; the way in which the TRC still only had *public* meaning through its "transmission by broadcasting" (Scannell 1995: 154), the orientation of the TRC towards media in its production and the myriad of ways in which what *became* the TRC for millions of people was an event which was mediated - *performed* on and by media, and therefore constructed through *media* narratives, conventions and genres. If these factors help to define a media event, as at times in Dayan and Katz's analysis seems to be the case, then the broadcasting of the *entire* process, not just the first week, must be seen as the media event, here a combination of the public ceremonial and the public political process, both mediatised. This of course means that we must also consider the possibility that media events can be seen as events that are more extended than the rather contained occasions described by Dayan and Katz, and that they are characterised by *clusters* of generic forms. Also, we must countenance the possibility that media events, like the TRC, could be oriented towards change and transformation rather than "order and its restoration".

Dayan and Katz introduce one category of media event, the shamanic, which goes beyond the celebration of order to raise the possibility of transformation. This category can form a starting point from which to think of the TRC within the broader notion of media event developing in the debates. Shamanising media events are those where the event itself offers the possibility of transformation, by introducing, enacting and giving voice to a new paradigm for society. Indeed, they argue that "the ceremonial broadcast of a proposal for change can, under certain conditions, actually induce such change" (1992: 145). They particularly have in mind events such as Sadat's visit to Israel, where the televised ceremony itself becomes the event, rather than addressing a prior event,

> Seeking neither to divest nor to restore, "transformative" ceremonies give full-dress treatment to a proposal for radical change, mobilizing an entire society to consider its acceptability (1992: 153).

Shamanising media events thus do not just halt the everyday; they do not just interrupt daily routines but actually disrupt taken-for-granted ideologies and paradigms of societies themselves. They force new ways of looking and thinking from within their own performance. For Dayan and Katz, such ceremonies can "beget and give meaning to new historical realities". They continue:

> The process involved is one of replacement, inside the ceremonial frame, of a given paradigm by a new one, and presentation of the new paradigm to the world (1992: 160).

This power of shamanising media events is not just that they present these new paradigms or models, they can actually bring about change as audiences activate latent or silenced aspirations to take up the new paradigm, challenging the old. History begins again, argue Dayan and Katz (1992: 164), as the media event marks a break, or an interruption, in the social order itself. Here, Becker's notion that mediatised public rituals broaden the base of reflexivity (1995: 642) is relevant, as the media form of the event greatly extends the possibility of reflecting on the significant of the cultural performance of rupture and transformation, from only those present to those unable to be present.[30]

South Africa's TRC had most of the elements of a media event as defined by Dayan and Katz, particularly in its shamanised form. As an event it was sponsored by elites, including the state itelf (1992: 153). It was designed by the Commission to be a ritual (Tutu 1999), carefully crafted to heal individuals but also to transform the country; to make the transition between old and new (Scheper-Hughes 1998: 14-19) and provide a new basis for unity. Confronting the past was very much about providing a new model, a new moral basis for individuals and institutions but especially for the *nation*. While the rhetoric of healing implied restoration, this was not restoration of the old order, apartheid, but rather of humanity and human dignity. If any older order was being appealed to in the ceremony itself, it was a version of the African humanist philosophy or world-view of *ubuntu*,[31] set in a new space and

---

[30] In this regard, Scannell (1995: 153) warns against seeing audience *of* the event in terms of a collectivity, certainly in the same way that those *in* the event are. He notes the possibility that media like television may domesticate exactly the qualities of specialness and awe which characterise public events, producing different ways of thinking about them in audiences at home (see also Becker 1995: 629). Although I agree with Scannell that one must countenance different perspectives for differently located audiences, surely Becker's point that reflexive possibility is extended through media access to an event is also well taken. Certainly in South Africa, people were able to reflect upon the TRC as a process and a form of imagining of the nation by engaging with the media performance of it.

[31] Tutu makes this plain in his published account of the TRC (1999), where he argues that *ubuntu* provided what he calls a "third way" between blanket amnesty and legal justice for perpetrators. He explains *ubuntu* thus: It speaks of the very essence of being human. When we want to give high praise to someone we say, '*Yu, u nobuntu*': 'Hey, he or she has *ubuntu.*' This means they are generous, hospitable, friendly, caring and compassionate. They share what they have. It also means my humanity is caught up, is inextricably bound up, in theirs. We belong in a bundle of life. We say,'a person is a person through other people'. It is not 'I think therefore I am'. Is says rather: 'I am human because I belong.' I participate, I share. A person with *ubuntu* is open and available to others, affirming of others, does not feel threatened that others are able and good; for he or she has a proper self assurance from knowing that he or she belongs in a greater whole and is diminished when others are humiliated or diminished, when others are tortured or oppressed, or treated as if they were less than who they are (1999: 34-5; see also Battle 2000).

time. And as an event, the ceremony *itself* was designed to be transformative, through its enactment of human dignity and connectedness in the form of eliciting narratives of the past which had been denied or suppressed. Yet it was the nation itself which the TRC sought to heal and transform, to bring into being in a new form, and this wider goal depended entirely on the TRC being a media event, a significant process which reached the public through the media. Through the notion of healing, the nation is the subject of the media TRC, interpellated (in the Althusserian sense) in its processes and representations, so that an entity which is in process is treated as already present, even if wounded. In the TRC media event the nation bears strong resemblance to Renan's notion of the nation as a soul (Malkki 1997: 67) and certainly 'the people' are constituted in terms of a moral community.

Opening up the concept of media events allows us to note that the media coverage which presented the public face of the TRC included *all* popular media, radio, television and print of all kinds. The TRC also had an official Website which published all briefings and press releases.[32] This diversity was appropriate in a country with a highly fragmented audience, high levels of illiteracy and large populations living in townships and rural areas without electricity. Spatially, the media field of the TRC was not confined to national and local media. Particularly when high profile cases, like that of Winnie Madizikela-Mandela, were before the TRC, its purchase became international. At various stages in her account Krog speaks critically of the 'intrusion' of foreign journalists, with their superior equipment and a lack of understanding of many of the cultural and political dimensions of the TRC.[33] As a media event the TRC seems to have been *designed* as a national rather than global one, though the ritual outcome of bringing the nation into being was crucial for South Africa's operation on a global stage.

As a media event, a process which depended on media for its ceremonial and political effects, the TRC worked with the representatives of all media in its production; a core team from radio, television and print media taking on the task of routinely framing the actual proceedings of the TRC to re-present it to the country and the people. Some of their names almost became synonymous with it, since journalists carried the TRC into the public realm on a daily basis.[34] It also had its own media personnel.

The reporting of the TRC thus became a very special form of the relationship which many analysts note between journalists and political figures.[35] First, media personnel saw reporting the TRC as special, different from routine news (cf Becker 1995: 643). This seems

---

[32] The official website (http://www.truth.org.za) provided updates and press releases throughout the whole process, extending the politics of inclusion in the process internationally through its role in making information public. In 1999 there were some problems (also around issues of public access to information) over the publication of the final report which involved the Website. The Webmaster has since begun a new website for "public information on the Truth and Reconciliation Commission" (http://www.struth.org.za/).

[33] Interestingly in 1994 there was also criticism of foreign media's handling of the first democratic elections. In particular the charge was again one of incomprehension:SABC-TV reporters in the field were frequently awful, but they were sufficiently hard working to bring an affecting picture of the election to viewers. Not as slick as the Allen Pizzey's or Peter Arnett's (both of CNN) of the world, they nevertheless had a better understanding of what the election meant to those taking part in it - and they conveyed that (*Combroad*, June 1994: 6).

[34] Antjie Samuel, who headed the SABC's radio team, recalls how anti-TRC Afrikaners in Ladysmith reviled her, the SABC and the TRC, all in the same frame (Krog 1998: 284,5) and how journalists received hate mail precisely because of their public association with the Commission, through their reporting of it.

[35] See Ward 1995, Hall et al 1978, Bourdieu 1996. Krog implies a mutually supportive emotional relationship too, when she argues that they all became part of the "same family" (1998).

to have enormously heightened their sense of involvement as did the rigours of the reporting process itself (Thloloe 1998). As the above quotations show, some journalists took seriously the possibility of media extending the reflexive base of the TRC, identifying a special role for media in bringing the TRC into the homes of the people in ways which directly fed into the Commission's postcolonial purpose. Du Preez argues that this was a quite conscious process, not just for journalists, but also for those testifying:

> For the first time, the nation acknowledged the victims. They told us that when they gave their evidence, they knew they were not talking just to the commissioners, but they were talking to *the whole nation*. That was the impact of the TV coverage (Thloloe 1998: 2, italics mine).

So media made the full ceremonial and political impact possible.

As a mark of this special media role, and the distinction between reporting the TRC and 'normal news', journalists held workshops to think about how to report it and what their role should/could be. Krog reports that they drew on the experience of German and Chilean journalists (1998: 19). Along with the tough, impossible questions of how one could keep one's own past out of reporting the TRC (so difficult for *all* journalists who had lived through apartheid), they had addressed the need to construct the media TRC as a *public event*, especially the issue of reflexivity, that is, how to involve the people in it. Always an issue for any media event,[36] if the TRC was to be conducted on behalf of the people of South Africa so that it really could contribute to transformation and provide the basis of a new national beginning, then this was a particularly crucial issue for the TRC as media event.

To this end, journalists and editors utilised and augmented all of the techniques and strategies of news and current affairs to produce a mediated TRC in which people could have a place, as involved audience, even if they were not participating directly. In response to the allied question of discouraging people from *refusing* the media event, SABC news bulletins made the TRC part of their routine content, using sound bites, narrative and drama to provide an engaging account of a victim's testimony in a few sentences and the press used headlines and photographs to catch the eye and compel attention even as they framed the event within particular interpretations.

There were consequences for this level of media involvement in the event; consequences for its co-production role. The TRC inexorably drew journalists and other personnel into the process, so that in a hyperreal way, the media covering the TRC became part of the story being told *in* media. For example, the television news of the first Gauteng hearings (SABC 1, 29.4.96) included a story of the difficulties and stresses for the media teams, and in print, journalists wrote themselves into a significant number of, even hard news, stories.[37] The Commission made available to the journalists the same counsellors that their personnel

[36] Dayan and Katz 1992, (Chapter 5) argue that a media event will fail if the public refuse to participate in their role of audience at home. This failure is exactly what occupied journalists as they planned their strategies for reporting the TRC.

[37] This was noticeable during the human rights violations hearings where some journalists confessed to being overcome by those giving testimony. See for example "I have never seen my husband's grave" (*Weekend Argus* 20-22.4.96) and "A bargain with the devil" (*Mail and Guardian* 20.3.97). In the special Christmas edition of the *Mail and Guardian* of 1996, Krog wrote "I am not made to report on the Truth and Reconciliation Commission" and went on to speak of her exhaustion as she became more and more haunted by the voices of the people coming to the Commission through the year. She reproduced and elaborated on this piece in her account of the Commission, *Country of My Skull* (1998: 50-51).

routinely used and journalists were able to develop techniques to deal with the stress, though Joe Thloloe, then Editor in Chief of SABC Television News says that it was not always accepted.[38]

There was another side to this, which exactly conforms to Dayan and Katz's argument that media's presentation of ceremony in media events is non-critical, appropriating the tone and purpose of the organisers. Journalists' commitment to the ritual purpose of the TRC affected the degree of interrogation that they were prepared to engage in as cultural brokers. Rather their brokerage role became reflection and interpretation and worked to stabilise the process rather than interrogate or challenge it (see Hamber 1997a). Though media reports did raise issues critical of events within the process, in an interview du Preez says that they avoided sensationalist reportage of intra-TRC problems (evidence of such disputes and other problems did appear but never became the big story even in more critical black and Afrikaner papers[39]). As cultural brokers, journalists had penetrated the field of the TRC to such an extent that they often appropriated its narratives and internalised its purpose.[40] At times, and quite appropriately so, they acted almost as an adjunct to the TRC, a kind of rallying ground for its cause. For example, SABC 2 made their current affairs debate program *Focus* available to Deputy Chairperson Alex Boraine for a highly publicised plea to the Inkarta Freedom Party to cooperate with the TRC.[41] All this is not to say that journalists did not raise critical questions about the TRC, we saw earlier that they did. But, by the time the Commission began to sit, many personnel had accepted its legitimacy and its value, which is hardly surprising given that media had already become the largest public arena in which the TRC took form. The media also now became caught up in the day to day job of presenting and representing the TRC's quest for 'truth', which coincided with journalists' professional ideals of getting the truth out to the people for them to draw their own conclusions. It also meant that they were drawn into the process.

The TRC hearings began in April 1996. As indicated earlier, these first hearings were given full, live media coverage on television and radio. After that first full five day television coverage, costs forced the SABC to limit coverage to daily news items and a weekly one-hour special report, which was broadcast on Sunday nights. By the time of the amnesty hearings, the TRC was routinely being covered in one and a half minute segments in the nightly news. Radio continued to broadcast the human rights violations hearings live. The difference was money available; where international funding had been raised for the radio broadcasts, Thloloe reports that the SABC just could not raise the advertising necessary to fund total television coverage. "No business wanted to see its name sandwiched between vivid descriptions of torture and murder" (Thloloe 1998: 2). They continued to tape the hearings as a record for the National Archives in Pretoria, but it was left to radio to broadcast the full

---

[38] Thloloe reports that the health consequences of covering the TRC intensively were often extreme. Max du Preez had two heart attacks in 1998 and another member of the television team suffered a breakdown, from which he had still not recovered (1998: 2).

[39] Tutu confirms this in his own account (1999: 157-9).

[40] The extent to which this purpose should directly become that of media reporting of the TRC did remain an issue however. In 1997, at a workshop organised by the Media Peace Centre, held in Cape Townjournalists and others debated whether the press should take an overtly reconciliatory role. While the dependence of the TRC on media to magnify its profile and proceedings on a national scale was recognised, it was also argued that media needed to be non-aligned with the TRC, in order to take a critical role if necessary (Braude 1997).

[41] The ANC also used media to encourage its cadres to participate. See "Go to the tc, ANC tells members" (*Mail and Guardian* 15.3.97).

proceedings of the event to those unable to attend; the ritual frame of each day at the TRC with its prayers and lighting of the candle; the welcome; the introduction and swearing in of each new person appearing; the questioning and prompting by Commissioners; the stumbles, pauses, repetitions and long breaks as people were unable to continue, often weeping and needing consolation. Together, radio, television and the press worked to mediatise this unique and history making political process, so constructing it as an extended media event.

Conclusion

A great deal of the work done on the relationship between nation and narration relies on the role of literature in providing both the pedagogic and performative registers for imagining nation. The TRC, along with other public events that are also media events, demonstrates a shift in the process of imagining nation, from the field of literature to that of media. Many things change when the form of publicly imagining the nation shifts, from print to particularly electronic media, from high culture to popular cultural forms. First, as theorists of media events observe and agents of nationalism appreciate, mediated national imaginings can reach an unprecedented number of people, regardless of literacy, wealth or station. Media events offered the possibility of bringing the ceremonial and political process of national imagination into the space of the everyday, into people's own homes. Thus, in the media TRC is found the pedagogic register of national imagining, a form that is authorised by the state, though not directly produced by it, is extended in manifold ways beyond the halls and committee rooms of the actual physical process.

At the same time, media are themselves performative. As a media event, South Africa's TRC was performed in, by and for media of various kinds. Media's performance practices, conventions and constraints intervened in its ritual and political processes to construct the TRC in ways which conform to media interests and needs, even where these were encompassed by, or reinforced, the intentions and needs of the TRC as a process. The TRC as media event provides us with an interesting intermeshing of the pedagogic and performance registers of national imagining. Certainly the most public version of the new history generated by the TRC as a basis for the new moral order of the nation, was very much an outcome of the media performance of the TRC. Since it was this which people appropriated and engaged in sites all over the country, we need to look further at the media performance of national pedagogic imagining.

In: Advances in Communications and Media Research, Vol. 3 ISBN 1-60021-188-7
Editor: Patula E. Grossmark, pp. 43-60

Chapter 3

THE NEW SOUTH AFRICA: REPRESENTING NATION AND STATE

Charmane McEachern
Adelaide University, Adelaide, Australia

The last decade of the twentieth century saw enormous interest in issues and problems of nation and nationalism, fuelled by competing movements. On the one hand, the breakup of the Soviet Union produced a proliferation of what Anderson (1996: 8) calls "congeries of weak, economically fragile nation-states". At exactly the same moment, globalisation with its transnational systems of production, communication and finance and the attendant routinisation of population displacement and migration, demanded a reappraisal of the role and meaning of nation and nationalism.[1] At the moment of movement from twentieth to twenty first century these issues seem somehow more poignant. Which form of collectivity will best suit the next century? Who will constitute 'we, the people'? Will nationalism and the nation-state continue to serve? National identity has become a major area of concern, particularly as it seems to be portable, no longer irrevocably coupled to territory (Anderson 1996; Malkki 1997, 1995), but part of a person which goes where he/she goes. In this context, South Africa's nation building quest is fascinating; it is a project to construct the nation as a territorially grounded unity at exactly the same time that globalisation encourages other kinds of collective identities, ones which transcend national boundaries. In such circumstances media in particular can operate paradoxically. They can contribute to these other identity processes and simultaneously generate national visions and imaginings which then become available on a global stage, projecting a desired image of South Africa and South Africans outside of the nation's boundaries as well as within.

South African imaginings of nation post-apartheid, in media and elsewhere, are deeply grounded in colonial experience, especially (but not only) in the apartheid past. As such, an account of South Africa's nation building processes, including the role of media, fits clearly into the study of postcolonial society, understood either in its minimalist, historical sense as societies which have emerged from periods of colonial rule, though forever in the position of

---

[1] See Appadurai 1990; Featherstone 1990; Miyoshi 1993; Held and McGrew 1993 Balakrishran 1996; Verdery 1991, 1996; LiPuma 1997.

having to respond to colonialism (Williams and Chrisman 1992) or in more complex terms as involving resistance to colonial processes, including the colonisation of the imagination (Spurr 1993; Chatterjee 1996; Werbner 1996: 3-4).[2] The relationship between state and media, the role of media in providing a unifying site for nation building and the particularities of the South African imagining of nation, all owe a great deal to the social forces which preceding forms of colonisation produced. Narration, memory and the past operate at the centre of South Africa's imagining of nation precisely *because* of this colonial past. So too does the attendant project to re-imagine space as part of nation building, which is the focus of the case studies of this book. Just what nation can be, its very possibility and desirability, how it is to be imagined, and who is going to generate such processes are all very much constrained and enabled by what has gone before. Building a nation in South Africa is a postcolonial project with goals of overcoming the exploitative and incomplete outcomes of previous versions of nation and society; especially the fragmentation and lack of inclusivity which ultimately came to characterise South Africa. Understanding the nation building process, especially the role of state and media thus first requires a discussion of that fragmentation and exclusion and the many social contradictions which these entailed. This takes us into the very relationship between nation and modernism, which is where this chapter will begin in order to place the nature of the nation building project and those agents who were most central in it.

Apartheid and Modernisation

South Africa's history of colonisation has been very complex, producing successive waves of settlement and displacement throughout the country. Particularly from its British phase in the nineteenth century, colonialism was an inherently modernist project, one which must be seen as operating coterminously with the development of capitalism. What became important for apartheid and hence South Africa today, is that, unlike other settler societies, colonialism was always incomplete, fragmentary and overdetermined. Even the question of the colonised territory was unsettled, upset and destabilised, often as a *response* to colonialism. This fluidity was not marginal, just a frontier problem. Suppression of border and other colonised populations for example produced massive, destablising migrations into the cities. Territorial incompleteness was an *inherent* feature of the social formation, an instability and a failure to achieve closure.

In South Africa, the modernist project had three major phases; starting in the pre-twentieth century stage (large and complex), moving into the period up to World War Two and then into the era of the apartheid state. Throughout, the *continuities* are as important as the discontinuities (Mamdani 1996). In particular, one can see that apartheid was the

[2] Ashcroft combines both of these senses, establishing a project which is far more than an historical category. It is far more than "after" colonialism, he says, not a stage in history at all. Rather: It begins when the colonisers arrive and doesn't finish when they go home. In that sense, post-colonial analysis examines the full range of responses to colonialism, from absolute complicity to violent rebellion and all variations in between. All these may exist in a single society, so the term 'post-colonial society' does not mean an historical left over of colonialism, but a society continuously responding in all its myriad ways to the experience of colonial contact (quoted in Ahluwalia 1997: 2); see McClintock 1993 and Shohat 1992 for critiques of postcolonial conceptualisations).

rationalist, inherently modernist, response to the incompleteness of preceding periods; response to the failure to achieve closure, itself a bid for closure (cf Mabin 1992b: 17).

Apartheid was a specific form of society, structured by social, political and spatial rules about race and the relations between races, enshrining the dominance and privilege of whites. Officially ushered in by the narrow electoral victory of the reconstructed National Party in 1948, in many ways it built on past modernisation processes, legacies from British and pre-apartheid, Union governments. In particular, it responded to the already existing relationship between urbanisation and modernisation in South Africa;[3] the growth of cities and the concommitant movement of populations from rural areas into urban ones provide the ground on which apartheid emerged and developed its policies. As such, the apartheid city is, as Mabin (1992b: 15) asserts, more than simply the "mapping of white political power onto the country". A dominant trope in apartheid as a vision of the world, the city also stood for modernisation as a form of economics and transformation.[4]

As many writers record,[5] the National Party was in many ways able to come to power with apartheid as its declared platform, precisely *because* of the strains which urbanisation and labour migration had already imposed on South African society. In other words, it was its rationalist proposals to address the lack of closure in this settler society, which allowed it to gain power (however slim its majority). The expanding and modernising economy was creating employment opportunities in the cities at the same time that mechanisation was reducing these in the countryside (especially after the Second World War), resulting in serious rural dispossession and urban migration (Mabin 1992b: 22). The National Party thus inherited a situation in which modernisation had produced cities out of control in planning terms, with fringe informal settlements and often grossly overcrowded, neglected and underserviced inner city areas.[6] As Mabin (1992b: 17) says, "In some respects apartheid was a (racist) response to previous failure to develop coherent urbanization policy". White complicity in apartheid at least partly lay in the economic benefits which accrued to them, of modernisation processes encouraged and protected by the state, particularly during the

[3] For accounts of this relationship see Maylam 1995; Smith 1992; Christopher 1994.

[4] As de Certeau recognises (1984), the city is the most powerful symbol of modernist conceptions of political economy, gathering to itself the tropes of progress and development as well as industrialisation (see also Harvey 1989; Jacobs 1995).

[5] See for example Worden 1994; Johnson 1988; Trapido and Marks 1987; Smith 1992; Christopher 1994.

[6] Cape Town is a good example here. Vivian Bickford-Smith (1995a, 1990) demonstrates the population build-up in inner Cape Town areas like District Six from the 1850s when the post-emancipation period saw ex-slaves rehousing in areas which quickly became slums (see also Ross 1992: 45). "By the end of the nineteenth century District Six was clearly established as a sizeable, residential area with a cosmopolitan population of approximately 30,000", reports Bickford-Smith (1990: 35). At the same time, with most housing being owned by absentee landlords and serious and continuous neglect from the Cape Town Municipal Council in an effort to keep rates low, the District also presented serious health problems. Of the Council Bickford-Smith (1990: 39) says: If they thought about the latter [District Six] at all this was usually the result of conditions in the area being revealed by a particularly severe epidemic. Thus that smallpox epidemic of 1882 produced a flurry of press reports on District Six and a temporary flurry of council activity aimed at improving at least sanitary conditions. This meant in practice some cleaning of streets, gutters and houses for the duration of the epidemic. But, as one Councillor conceded, most streets were not made up, lacked guttering and water, let alone subterranean drains and were pockmarked by "pools of black filth in every direction." Overcrowding with all of its attendant problems was exacerbated as the twentieth century began and rural populations moved in, attracted by the possibility of work at the nearby docks (though workers here also lived in on-site hostel accommodation which imitated labour conditions in the mines up north).

nineteen sixties when the apartheid state was transforming and building its new society, with at least the outward appearance of minimal resistance.[7]

The pre-apartheid situation also contained the seeds of Mabin's "racist response". As problems, the above strains were already being interpreted in oppositional, racial terms. At the same time there was already in place a system of Native Reserves, locations, pass laws and some influx controls, all of which were supposed to control immigration to the cities but which were seldom enforced with any rigour until apartheid.[8] Segregation had been debated as desirable by many, often even for liberal reasons, but as Worden reports, this was largely breaking down in the late 1940s, part of the failure to achieve closure. Race though *was* entrenched as a major contradiction, overdetermined by the articulation of pseudo-scientific theory emerging in the nineteenth century (Dubow 1987), economic and social policy and development and modernism itself. As such, race was available as a structure and a discourse for the modernising solutions of the new apartheid state, providing the dark side of the enlightenment even as it was put forward as a solution. Huyssen (1995) argues that one essential aspect of modernism is always its *anamnesis*, its intolerance of diversity,[9] seeing in all forms of colonialism the manifestation of exactly this vision of the world. In South Africa now, it provided a focus on space and division deep in the modernisation project.

Apartheid, as it developed through the fifties and sixties thus went further than mere urban development, as it mapped racial constructions onto space in a thorough going way. Although, as Mabin (1992b) argues, apartheid policies were often more reactive to the situation than ideological, certainly in any unified way, the technocratic and rationalist solutions of the apartheid state came to focus on race. Racial segregation itself became the goal of apartheid, its solution to the problem of closure, and a far more systematic and comprehensive racialised system for modernising South Africa. The Population Registration Act of 1950 was the lynch-pin,[10] and the 1960s, apartheid's most ascendant period, were marked by a myriad of petty and bureaucratic laws and procedures to 'perfect' the system of racial separation (P. du Toit 1995). Africans were defined as '*foreign*' elements in the cities and a whole range of policies were mounted on a series of antinomies which not only defined people dominantly in terms of race, but also hierarchised them; rural /urban, civilised/ primitive. Bantustans, later homelands, developed on Native Reserves, were to be the vehicles for enacting and facilitating the separate development of Africans and tens of thousands were relocated into them as their nationals (thus losing their South African national identity).[11] Henceforth, Africans (as well as those classified Coloured and Indian) were to live and work in the cities on the state's terms. Modernisation depended upon their labour, but its benefits were distributed on a different set of principles. As Smith et al (1992) note, this contradiction

---

[7] For accounts of this period see Johnson 1988; Worden 1994; cf P. duToit 1995: 329-30.

[8] See Mamdani 1996; Worden 1994; Trapido and Marks 1987 for accounts of pre-apartheid forms of racialised population control.

[9] Nairn (1997) points out the paradoxical nature of this dimension of modernism as it gives rise to the nation-state. He notes that in fact the recognition of diversity is the condition on which the nation-state is built.

[10] This Act made possible the panoply of acts necessary to structure segregation in all areas of social life, in particular the Groups Areas Act which was central to apartheid ideology and rule (cf Maharaj 1992). Although many of the legislative acts which enacted and supported apartheid as a particular system of rule were repealed during the nineteen eighties as apartheid attempted reform, the Population Registration Act existed until 1991 (P. du Toit 1995: 375).

[11] As Worden (1994) observes, this necessitated totally ignoring the inability of the old "reserves" to maintain such numbers of people even at subsistence level.

sowed the seeds of internal strain in the system itself.[12] It produced an irrevocable problem for apartheid's efforts to achieve completeness for its modernist project.

The apartheid state was thus essentially a modernist state, drawing on the legacy of past governments, both colonial and those of the first postcolonial phase in South Africa's history, but greatly extending these through its particular form of racialised solution. As a modernist project, apartheid was committed to (separate) development, economic growth and forms of improved (if racialised and restricted) prosperity, and classification, separation, division and measurement were among a whole range of modernist and enlightenment social tropes which characterised it.

Also entailed was a modernist project of nation building, itself part of the state's attempts at closure, though ultimately this too was unstable and contradictory. On the one hand the state promoted white nationalism as a strategy to secure white cooperation for and identification with a South African nation defined primarily as white.[13] At the same time there were radical, ethnicised differences among whites which threatened nation building. In fact, some of the particularities of the apartheid state can be explained by a contradictory anti-British nationalism (Krog 1999;[14] Goodwin and Schiff 1995: 29, 67; 187-188) which was directed at re-ordering the relationship between British and Dutch colonialism and enshrining the political dominance of the Afrikaner on the Anglo South African.[15] When imposed on an inherited fragmentary structure which became increasingly intolerant to non-European difference, this gave South Africa a strange character of being simultaneously postcolonial and colonial; defining itself in opposition to a past, oppressive British colonial presence, while at the same time operating a system of internal colonialism on the basis of racialised difference; anamnesis.[16] In this regard, one could quite easily use McClintock's careful exploration of the different outcomes of colonisation globally to characterise apartheid South Africa as a system of internal colonisation by a "break-away settler colony" (1993: 295). This Anglo/Afrikaner division among whites remained a fault line throughout (du Toit: 354). It was symbolised in the language debate over the relative status of English and Afrikaans, which in turn became one of the key points exacerbating tension between the black majority and the apartheid state. It operated as a further tension because it was an 'ethnic state' that created nation based on racial separation.

The state/society line becomes very shadowy and broken here, partly *because* South Africa's modernist project was always unstable, with its actions always intensifying the internal divisions rather than healing them and producing completeness (Simpson 1994:

---

[12] De Villiers (1987: 369) argues that the Western Cape was probably the best example of the "clash" between the politics of apartheid separation and the social reality of distribution, since the social relations of the Cape derived from "three hundred years of mixing" among the various populations.

[13] Here, Simpson (1994: 470) for example, talks about the role of the border wars in this process of unifying whites, despite differences between Afrikaans and English speakers, while Booth (1996) investigates the contribution of sport.

[14] Krog recounts how fundamental the Anglo-Boer War was to the myths of the Afrikaner nation, transformed into nation building narratives of The War, an epic of bravery and betrayal which in turn provided a manifesto for living.

[15] This is why du Toit argues that the regime must be characterised through the concept of an ethnic state (1995: 331) and that the particularities of the segregationist system can only be understood as ethnic politics.

[16] In this context, it is noteworthy that the Bantustans were established within the rhetoric of decolonisalisation (Johnson 1988), in keeping with the emergence of independent, postcolonial states in the rest of Africa of this time.

466). The nation as it was officially constructed under apartheid is a primary example of this striving for completeness; the state used separatism as its device for asserting the unity which Hobsbawm (1990) argues the modern nation-state always seeks.[17] As a form of territory and identity specifically generated to dominate, exclude and disenfranchise the non-white majority, such a nation was also a prime example of the imposition of nation-states in Africa, as products of colonial divisions (Chatterjee 1996; Williams and Chrisman 1992).[18] Along with this went large areas of law-free operation and coercion, enabled by the restriction of suffrage which forced a huge percentage of the population into the status of the *objects* of rule not the *subjects* as in a modern nation-state. None of this could accomodate the needs of modernisation as discussed earlier.

These social projects and their attendant forms of modernism thus failed. The society did not grow. Forced to direct huge amounts of its resources and effort into maintaining state and society against the weight of its own contradictions as well as global protest in the form of sanctions, it stagnated and imploded (P. du Toit 1995). Thus, the apartheid state also failed to achieve closure though it constantly tried to do so, modifying and reinventing itself over and over again. Though, from the outside it may have had the appearance of a total system, it was always fighting for closure. Ultimately, in the eighties, the system tried first to bend to accomodate these strains (orderly urbanisation) and finally ended in this form. In the long run, modernisation destabilised apartheid's project, which itself was causing instability through the intensification and embedding of division. At the same time, the completion of modernisation also was imperilled, internally and externally by what apartheid had become. It did not serve modernisation to have illiterate workers (Breuilly 1982), particularly as capitalism moved into its late phase. It did not serve modernisation to have workers illegally camped on the edge of the cities. It did not serve modernisation that apartheid prevented South Africa from moving fully into the global economy, because of sanctions. Modernisation in South Africa was further continuously destabilised by its concommitant impossibility in the inadequately resourced homelands (Butler et al 1977; Christopher 1994). The very separatist structures designed to aid in the task, in the end subverted the modernist project itself. Apartheid was no longer compatible with the needs of capitalism, the apartheid version of nation and state becoming encumbrances rather than strengths at this stage in history.

With the collapse of apartheid and the rise of an ANC dominated African democratic order, the modernisation, developmental and national project could be restarted and reinstituted. Here too though we find the influence of the legacies from the past. Central to the new version of the modernisation project was the need to build a common or shared understanding of what South Africa was and what is could be.

---

[17] This exclusive strategy and definition of the nation was a major problem for the National Party when it first came to reconstruct itself after apartheid, as the New National Party, with the need to find more a more inclusive vision of nation which at the same time provided continuity in its form of identification. Initial strategies at least used language with the Party extending its vision to all Afrikaans speakers.

[18] The homeland of Bophutatswana for example was constructed out of a series of fragments, separated by 'South African' buffers and having no uniting infrastructure (Christopher 1994; Worden 1994).

Modernism and the Post-Apartheid State

In 1994, the legacy from the past in South Africa meant that the only way that modernisation (and modernism) could be completed was through a post-apartheid state. Colonialism could not achieve closure, nor could apartheid's version of internal colonialism and its legacy of division and fragmentation. What is significant in the light of what has gone before, is that, for this inheritor of the modernist project, modernism is being conceived in terms of unity *and diversity*. The conception of modernist unity here is not grounded in or enacting of *anamnesis*, fear of diversity, but rather its opposite and the task for the new state in its nation building project is the harmonising of diverse cultures. Modernism in this emerging situation then is crucially coupled to, and dependent upon, a totally different vision of nation and state, one which necessarily carried within itself both its past and its visions of the future. These temporal constructions were themselves encompassed by implicit modernist critiques of South Africa's history so far, while the very diversity which they sought to accomodate itself owed much to past, apartheid visions of peoples, nations and state. Clearly, we need to look more closely at the process of nation building in the new South Africa, for how it seeks to further the modernisation project by overcoming the problems of its predecessors.

Breuilly (1982, 1996) argues that nationalism is always essentially political.[19] He criticises Gellner's transformative model in which nationalism is an essential element of the transformation from an agrarian based society to industrialised one, and argues that there is no singular path to nationalism. Verdery (1996) also cautions us about the traps in theorising nation and nationalism, particularly in postcolonial societies where there is a danger of talking about one thing using concepts which apply to another. We need to understand exactly what kind of nationalism is being constructed and appealed to in the discourses being deployed at a particular time and place. How is the nation being constructed? By whom and for what purpose? In whose interests? Here, we must take seriously the symbolic construction of nation, as Verdery insists.

In this regard, Verdery uses the work of Brackette Williams (1989) to suggest a generative role for the state in the process of nation building (cf Alonso 1994). In South Africa today, the state *is* central in the political processes constructing nation (P. du Toit 1995: 352).[20] As LiPuma (1997: 237) indicates, new states like that of the democratic South Africa *must* operate on a global stage, which overdetermines their choice of the nation (as in the nation-state) form, regardless of today's challenges to it. Despite the highly problematic legacy of a "national order of things" (Malkki 1995; 1996: 54) in which the concept of 'nation' as the norm for collective identity and belonging is tainted by apartheid's particular, racialised and ethnicised, exclusive version (Simpson 1994; Thornton 1996), the post-apartheid state too finds itself embarking on a nation building project (P. du Toit 1995: 352). Just *how* it is doing this is in turn influenced by the nation building strategies of the apartheid state.

---

19 See also Williams 1989; Verdery 1990; Mann 1996.

20 Contrast this view with those of Thornton (1996) and Giliomee (cited in du Toit 1995: 327). Giliomee, for example, develops a nation formulating state equation, using the history of Afrikaner experience in South Africa. Yet the accounts of du Toit and others indicate clearly that it was the apartheid state, dominated by Afrikaner interests, which actually took up a very particular kind of national imagining in a nation building project which suited its particular nature as an ethnic state.

In *its* nation building processes, the apartheid regime used the rhetorics of independence and postcoloniality (particularly as they operated in other parts of Africa) in order to legitimate its fragmentation of the territory of South Africa on a racial basis, declaring certain parts 'independent' states for Africans to 'develop separately' from white South Africans. In the sense that this was a state construction of South Africa as a nation for whites, it is the *post*-apartheid period which can be seen to bring in to being the kind of postcolonial, modernist (modernising) state which was the actual basis of the rhetoric they were using. As Hobsbawm says, the apartheid construction of nation(s) must be seen as "not an exercise in classical nation building for Africans, but as a project to perpetuate racial oppression" (1990: 158).

What the past history of nation building means is that the new state reigns over a fragmented society which various analysts have even described in terms of a constellation of different temporalities and spaces. Thornton (1996: 146) uses the newspaper coverage of a hostel murder to very good effect to demonstrate the enormous fragmentation and resulting confusion and play of identity in South Africa today. Just part of this reads:

> Some hostel residents told *The Star* the violence started when a Pedi man killed his Mosotho wife and a Mosotho man after finding out they were lovers.
> Media reports that the two were killed because they refused to contribute money to buy firearms for the hostel were nonsense, a group of women said.
> A Mosotho woman...said the two were killed after they were found having sex by the husband.
> A Basotho faction then concluded that the lovers were killed by Xhosa resients and retaliated.
> But other residents said the clashes were caused by AmaBhaca and Basotho residents who, together, with Inkarta Freedom Party members, were trying to remove Xhosa residents from a section of the hostel to make way for IFP members....

There is constant accomodation to this kind of fragmentation in state rhetorics and it presents a particular challenge in the construction of an emergent nation, but the ANC also opposed a solution in the form of a nation of minorities, the preferred NP option which was supported by the ethnic nationalism of the IFP.[21] For the ANC, *unity* and diversity are central features of their desired polity for South Africa, raising questions about just how *national* diversity might be imagined. What might be shared in such a nation? What symbolic construction can meaningfully combine diversity with unity? To these questions might be added also, where is any essential otherness to be found now? These could be seen as reformulations of Hobsbawm's ideas of transformations in nation-states at a time when such entities no longer have an historical mission (1990: Chapter 6).[22] Here, they raise the issue of just what conceptions of nation have been available to South Africa since 1994? Initially, the liberation movements and the struggle against apartheid provided at least *part* of the content of the imagined nation, as a collective identity transformed from the exclusions of apartheid to the inclusivity of the new.

---

[21] On this issue see P. du Toit 1995: 379; Muthien and Khosa 1998; Johnstone and Johnson 1997.

[22] Breuilly (1996: 171) also argues that, with some exceptions such as parts of Eastern Europe, nationalism is today disappearing as "a genuine politics" but that still exists as a "rhetoric employed by all national governments". Certainly, the politics represented by and within South African nation building as detailed in this chapter are quite different from that which he identifies, as one would expect in postcolonial Africa (cf Werbner 1996).

Historically, apartheid rule and the struggles against it generated two different uses of nationalism which have contributed to the deployment and conceptualisation of nation by the post-apartheid state. These notions of nation or national are informative of attempts to forge unity out of diversity in the name of nation. Hobsbawm (1990) has differentiated between nation understood to symbolise the claims of political rights within a polity - the rights of citizens, and nation based on ethnicity, that is some criteria of culture, language, foundation myth or shared history.[23] The apartheid state used the latter form, devolving from Afrikaners' own constructions of themselves as an ethno-nationality, a Broederbond inspired conscious political effort to forge unity among Afrikaners of different positions and different consciousnesses which dates from the nineteen thirties.[24] It was this view of nation that produced that fragmentation of South Africa and this is certainly *not* where the national claims of the various oppositional bodies like the African National Congress or the breakaway Indian National Congress were to be found. Rather they deployed the first, democratic, kind of meaning identified by Hobsbawm, where nation symbolises the claim of a majority of the disenfranchised; it is a claim for political rights.[25] It is this form of national unity which most strongly symbolised nation in the immediate post-apartheid period. In the post-apartheid state we consequently also find the kind of sensibility which is embedded in the liberal state, described so cogently by Poggi (1978) for the twentieth century state. Here too, the state is a form of rule, which functions "within and through political and juridical forms derived from the liberal-democratic nineteenth-century constitution" (Poggi 1978: 121; see also Orkin 1995). In South Africa, the new constitution is exactly a major juridical work of consultation and inclusion, giving form to another principle of the modern state, representation of/by the people, itself manifested in maximum enfranchisement.[26] In this sense the state itself provides the principle of unity necessary for nation.

In this political form of nation, diversity is nothing to fear because it is pushed into second place by *citizenship*.[27] It is as new *political* subjects that all South Africans belong to the nation and unity is signalled in the new South Africa by the post-apartheid state ruling through universal enfranchisement (contra apartheid). Imagined nationhood here has its origin myth in the liberation movements themselves, which resonates with Werbner's account of the same phenomenon in Zimbabwe:

> Some nationalists advocate the idea of the popular nationhood of the people's liberation war. Their cause is the homogeneous nation, the nation diffused throughout the country, and dedicated to the liberation ideals of freedom and equality for all (1998: 75).

---

[23] Giroux (1992) points out that these two are not mutually exclusive, using the right-wing backlash in the United States to demonstrate that the latter, 'cultural' form of nation may still be mobilised after the former, political form, has been secured. Hage (1998) explores something of the same thing happening within Australia's multiculturalism.

[24] For more discussion of this important aspect of apartheid see Goodwin and Schiff 1995; Hofmeyer 1987; Worden 1994; Breuilly 1982; Kinghorn 1994; Wilmsen et al 1994: 348-9; Krog 1999; de Villers 1987: 341; O'Meara 1996.

[25] See Breuilly 1996: 165; Giroux 1992: 4; Simpson 1994.

[26] Poggi argues that the drive to ever greater enfranchisement is a feature of the modern state (1978).

[27] LiPuma (1997: 234) also notes that in the Solomon Islands where the state had similar problems of contructing a uniting vision of nation in a situation of enormous diversity, citizen became the most powerful subjectivity.

In this sense, the new South Africa is postcolonial in that it overcomes that contradiction at the heart of colonialism noted by Comaroff; civilisation without citizenship (1998). This is the form of nation which Simpson (1994: 473) argues provides the greatest hope for South Africa; the 'civic nation', with enfranshisement providing the form of belonging that unites all; the people conceived as citizens. In this sense of nation, one of the most powerful symbols of the new South Africa was the massive queue outside election booths in 1994; political inclusivity as the symbol of the attainment of the modern liberal-democratic state form. Citizenship is also thus the most concrete form of the much vaunted unity in diversity, with the emphasis on *unity*; politics encompassing difference, social, cultural or economic, to produce a whole that is greater than, because *above*, ethnicity.

The South African state has clearly, consciously, embraced this political conceptualisation of nation as a manifestation of the ideal of unity in diversity. Consultation and the dissemination of information became primary concerns in all areas of society, from the Constitution and the elections to the Reconstruction and Development Plan (RDP). Not only did this emphasis flow from the democratic principles on which the new nation was to emerge; democracy depending on informed citizens, it also publicly encouraged and symbolised the inclusivity of this new nation. The media were central in this. In such a fragmented society the nation conceived as democratic inclusion has provided a crucial role in nation building for the media, particularly the electronic media, which, under state direction, were transformed and mobilised in the name of just this form of nation (Revill 1994; Teer-Tomaselli 1995).[28]

MEDIA, STATE AND NATION

Media of all kinds, with few exceptions, were deeply implicated in apartheid's rule, legitimating it and silencing those who were constructed as the objects of rule through a range of strategies of exclusion and othering.[29] Dominated by monopolies[30] and operating in a legal situation in which journalists and editors were frustrated by the fact that the public was deemed to have no democratic 'right to know' (*Between the Lines*, SABC, 19.5.99), complicit reportage of the regime's activities was also generated by media vehicles which were irrevocably involved with the ruling party. Afrikaner newspapers have most often been cited

---

[28] Louw for example makes this link clear in the Preface to his edited collection on the debates around South African media reform in the 1990s:

A society's media system is intimately connected with providing the basis, or otherwise, for democratic government. Democracy, in turn, impacts upon the possibilities people have to influence their society, and hence their own life-chances. For this reason the media policy debate, outlined in this book, needs to be taken seriously by South Africans as a central issue in the re-building of their country (1993: 4).

[29] See Phelan 1987; Burnett 2000; Louw 1993; Evans 1992.

[30] In the early nineteen nineties for example, Louw writes that five media institutions dominated the South African media: the South African Broadcasting Corporation, Argus Holdings Ltd., Times Media Limited; Perskor and Nasionale Pers Beperk (1993: 160). In turn, Argus, dominated early by mining-capital, was controlled by Anglo American/JCI (ibid 163) and also had significant interests in electronic media, especially M-Net television (ibid 165). Louw reports (1993: 163) that its mining interests caused the Argus Company to "see-saw", "back and forth from highly disapproving, to moderately unhappy, and occasionally approving - in its responses to NP rule as the interests of capital have been advantaged or disadvantaged by this Party's policies".

here (corporations like Nasionale Pers favourably treated by the regime (Jones 1998: 331),[31] but the English media was also involved; the Johannesburg daily, the *Citizen* was established with state funding for example. The South African Broadcasting Commission(SABC) also operated as a para-statal; so much so that its statutes and practices actually had to be changed in order for it to operate appropriately in the 1994 election campaign.[32] Particularly in the 1980s, the SABC 'explicitly' supported the government's policy of Total Strategy and overtly became part of its implementation (Teer-Tomaselli 1995: 578). Beyond actual complicity of course, was a range of routine self-censorship and editorial policies that favoured the accounts of the state and National Party Government.[33] And the application of apartheid racial discrimination in media occupations further limited the range of views available in any accounts. The complicity of media in apartheid was such that it became the subject of a special TRC hearing on the media in 1997 (Burnett 2000).

Of course there were also numerous acts of brave and principled defiance, of refusal to comply and efforts to achieve best journalistic practice, often costing the perpetrators dearly, especially during the 1980s' periods of emergency, when more extreme state censorship was also applied (Phelan 1987; Teer-Tomaselli 1995: 578).[34] The newspaper *New Nation* exploded the inquest findings into the deaths of the Pebco 3 in Eastern Cape by publishing a top secret military intelligence message which demanded their "permanent removal".[35] *Cape Times* journalist Tony Weaver was tried for the anti-apartheid story that he produced of the security police killing of seven youths in the township of Guguletu (the Guguletu Seven) (*Truth Commission Special Report*, SABC 2, 28.4.96). In the TRC hearings, journalists trying to do their job emerged as the subjects of the stories of gross human rights violations.[36] Obviously apartheid presented a very hostile and difficult environment to be negotiated by many media professionals.[37]

---

[31] Indeed, in 1997 *Die Burger*'s editor, Ebbe Dommisse admitted that the paper had been "the official mouthpiece of the [National] party in the Cape Province" ("'No longer NP mouthpieces'", *Mail and Guardian* 21.2.97-27.2.97: B6).

[32] See Teer-Tomaselli 1995; van Tonder 1993; Revill 1994; Fry 1994.

[33] A number of analysts have discussed these strategies, Phelan 1987; Louw 1993; Teer-Tomaselli 1995: 577-9; *The Media Report*, 13.6.96. Further, in the latter edition of Australian Radio National's *The Media Report*, producer Adrian Herring, who initially worked for the SABC, told of the editorial policy of systematically referring both scripts and completed programs "upwards", where factually correct, but politically sensitive programs (for example about racially based differences in education under apartheid) could even end up with as high as personage as the Director General of the SABC personally refusing to allow transmission.

[34] In the submission to the special media TRC that he wrote on behalf of Times Media Limited, Rex Gibson, formerly editor of the *Rand Daily Mail* and former deputy editor of *The Star*, expressed his concern that the many journalists who did try to report what was happening would be "crucified" in a general condemnation of all media personnel and institutions, one which failed to draw distinctions among journalists, media vehicles or between journalists and proprietors and managers (*Mail and Guardian* 25.7.97-31.7.97: Inside Opinion 26). He particularly indicted the "draconian legislation" which silenced journalists, asserting that, "Black journalists were often detained without trial in terms of the notorious detention laws, and sometimes tortured."

[35] Despite the evasions of FW de Klerk in the matter of the revelation of the Barend du Plessis minute concerning the killing of Matthew Goniwe, one of the Pebco 3, Truth Commission amnesty proceedings revealed that "permanent removal" was intelligence code for execution (Electronic *Mail and Guardian* 1.6.99).

[36] See "Detainee died after being beaten, tortured" (*Cape Times* 23.4.96: 3).

[37] Problems clearly still remain. In 2000 the Human Rights Commission published a report "denouncing the South African media as racist and calling, among other things, for racism awareness training for journalists" (Beresford 2000: 3). And du Venage (2001)reports a complex interrelationship between race, media and politics in politicians' attacks on journalists for criticising government, clearly a legacy from the past.

Despite the many and varied exceptions, media of all kinds would appear to have systematically assisted the state in its construction of legitimacy and consensus which meant that after 1994, the inclusive, political form of nation in which media have a central role to play in fostering democracy and citizenship required a major transformation of the media landscape. Media were consciously understood as crucial institutions in the new democracy, requiring shaping and organising for greatest inclusivity to assist in nation building programs. Although print media were recognised as best suited to the sophisticated development of ideas, television and radio were favoured as vehicles most able to get information to a mass audience. Popular genres, such as soap opera and sitcom were deliberated used by elites to promulgate detailed qualitative information about issues as diverse as elections and voter registration, the RDP, trade unions and health. The narrow, parochial focus of news under apartheid was transformed and transcended by the deliberate creation of a national network capable of involving the whole country – both through its signal coverage and its democratisation of internal organisation and work practices.[38] At the same time, strong emphasis on inclusion meant the employment of greater numbers of black journalists[39] and local language broadcasts (Teer-Tomaselli 1995: 590; van Tonder 1993). In radio, the most inclusive medium in these terms,[40] we can see parallels with Anderson's discussion of the relationship between nation building and the rise of vernacular print media (1992: Chapter 5). Radio broadcasts of the national elections, but in local community languages (there are eleven official languages in South Africa) drew people together and made possible a sense of belonging as members of a national political community.

This relationship between media and nation, theorised by Anderson, was quite deliberately and politically made in post-apartheid South Africa, influencing policy, organisation and regulatory institutions like the establishing of an Independent Broadcasting Authority.[41] In a speech to the World Summit on Television and Children in March 1995, Dr Ivy Matsepe-Casaburri, then Chairperson of the SABC, made the connection between media, politics and nationhood in terms of citizens quite clear when she said:

> The role of broadcasters, especially the public broadcaster(s), in anchoring and nurturing democracy and building the so-called "rainbow nation" cannot be and is not underestimated. This still dichotomised society has not only had a history of racial and gender oppression but also geographical fragmentation and subordination from which it needs to be freed or to escape (1995: 9).

---

[38] Just for the SABC, Bredin (1994: 29) notes the need for democratic on-air presentation and internal practice, training in "the new ways of democracy", the agreement of a new ethical code and "a statement of new values and new visions for a public broadcaster in circumstances of social and political upheaval", just to provide satisfactory election coverage.

[39] The notions too that institutions like the SABC required 'de-Afrikanerisation' also produced the employment of black announcers and hosts (see for example *Mail and Guardian* 1994, 10(1): 34.

[40] In its October to December 1993 edition, the SABC *Radio and TV* journal reported enormous growth in "black radio" (p. 34), with daily audiences of 17.6 million adults. African language broadcasting stations accounted for 11.2 million listeners (figures did not include Radio Metro, the national English language station aimed at black urban listeners). Interestingly, the link between radio, information and democracy is made when the article goes on to report the need at this time to improve black radio stations' reach to ensure parity between stations in time to provide adequate services for the upcoming elections. It makes the point that SABC radio had a "pivotal role in voter education", especially for black people voting for the first time.

[41] See for example *T.V. International* 1996: 5-8; Mtimde 2000.

So media, especially broadcasting media which best serve a nation with huge levels of illiteracy (Teer-Tomaselli 1995: 585), were again coupled to nation building in a fundamental way. This time their role was to support the programs of a postcolonial state, operating in the name of inclusivity and political education and participation, not exclusion and domination. In turn, the newly re-organised media, could be seen as symbolic of the desired new, emerging nation, at least in terms of their rhetorics of freedom of expression and information (Fallon 1996: 9).

Media were thus deeply implicated in the political form of nation. But their role went much further than this; they also provided one space in which visions of nation beyond the political sense were publicly demonstrated, elaborated, debated and contested. This meant that media figures like Matsepe-Casaburri became significant figures among those elites which began to produce and support state generated agendas of nation building. Matsepe-Casaburri again:

> But the concept of a "nation" has only just emerged and is not yet a reality. Another great challenge in South Africa is to build a sense of nationhood united around a common sense of purpose whilst recognising and celebrating our diversity of cultures, opinions and life experiences within the nation. We know broadcasting can play a role, if not the most important role, because we have seen how broadcasting played a role in promoting apartheid ideology and policies of separate development and cultural oppression. The democratisation of broadcasting, as illustrated by the Independent Broadcast Authority's mandate, is a bold beginning in allowing new forces to address the past (1995: 9).

Here, Matsepe-Casaburri evokes nation and links media to nation in crucial ways which, although they directly support state initiatives and visions, discourses and rhetorics, also extend the imagining of 'nation' and 'the people' beyond the democratic, political form of citizen. In their cultural and expressive work, media elaborated the original vision in ways that vastly extended the state's imagining even as they supported it. This meant that media could play a further role in national imagining when the inevitable legitimation problems of a postcolonial state began top be played out.

Media, Postmodern Society, Postcolonial State

The image of the election queue appeared in media all over the world, symbolising the new, political, nation. It also demonstrated the incompleteness with which the post-apartheid state had to contend. For the queues themselves (and the many problems in the election process such as the late entry of Inkarta Freedom Party and the menacing of candidates in areas where they had little or no support) revealed just how incomplete the modernisation process was and how much work remained.

History has not stood still and South Africa has moved on in ways which have increased and intensified the differences and tensions in society, rather than eradicating or containing them. The fragmentation which was historically such a barrier to the modernist project has not been dismantled with a new political regime. The attainment of citizenship and political rights *are* crucial aspects of nation building, part also of the ongoing reconciliation process in South Africa today (see chapter 2). But, as Alex Boraine insisted when he addressed the

National Reconciliation Council of Australia in Melbourne in May 1997, admitting what happened and accepting responsibility, cannot on its own be enough for a new South Africa. Reconciliation must be supported and legitimated with *real change*, which directly addresses the inequality embedded in the continuing fragmentation.

The *legitimation* of the state depends on the achievement of political rights being accompanied by economic and social rights, adequate housing, water, power, jobs and so on. Again, modernisation moves to centre-stage (Simpson 1994; Ahluwalia 1997), only now it is coupled with a democratic politics. It becomes *politically* defined as modernisation to deliver benefits to *all* citizens, not just a racial minority. What we see in South Africa then is the state trying to shape economics, at least in rhetorics, through politics. For example, the ANC's election slogan was "A Better Life for All" and the Reconstruction and Development Programme (1994) also exactly fitted this strategy, combining economic reform with an inclusive political agenda. Its very first item sums this up:

> The RDP is an integrated, coherent socio-economic policy framework. It seeks to mobilise our people and our country's resources toward the final eradication of apartheid and the building of a democratic, non-racial and non-sexist future (1994: 1).Writing of the Solomon Islands, LiPuma (1997: 230, 231) sees such national development programs as shaping "a nation that corresponds to the state. But "nation as development" has been problematic in South Africa.

South Africa's problems are not entirely new, corresponding to a situation identified by Poggi when he notes, following Weber, that as the modern state develops, it begins to feel the inadequacy of the legal-rational legitimacy on which it depends for the compliance of its citizens (1978: 101). Legalistic norms are very weak in terms of moral ideals that might bind people to them, so that the modern state has had to find some other form of justification and moral authority. For Poggi, the logical domination of economics in modernism (1978: 120) means that the administration of the economy itself takes that role.

> Particularly in the 1950's and 1960's, an ideal variously termed "industrial development," "economic growth," or "affluence" gained an overwhelming grip on the public imagination. It was unanimously endorsed (at any rate in their rhetoric) by political leaders of all persuasions, who treated it on the one hand as utterly self-justifying, and on the other as validating whatever burdens the state might impose on society (1978: 133).

This would seem to be the form of legitimation that most visibly followed or is entailed in the extension of enfranchisement in South Africa as it is constructing itself as a modern liberal-democratic state, despite its postcolonial context. The ANC ran on a platform in the 1994 and 1999 elections (and indeed the local elections of 1996) which was all about its capacity and will to modernise in the interests of all 'stake-holders' (itself a term of inclusivity which refers to state more than nation) (see Bertelsen 1995).

Poggi also talks of the serious problems facing a state and a political process that has reached this point. Even in societies *without* the history of partial development and serious inequality of South Africa, states are having trouble delivering affluence and development of the kind that would help them to maintain their legitimacy. What du Toit calls the weakening of society (1995) under apartheid, foreign debt and the failure of the global community to reinvest satisfactorily in South Africa are taking their toll. Economic and social transformation lags well behind its projected timetable in many respects and there are serious ramifications resulting from the tax revolts, investment boycotts and anti-apartheid activity of the 1980s (P. du Toit 1995). The issue of increased violence has become the synecdoche for this lag; made highly visible in media and construed as 'crisis' it becomes another form of the failure of closure, this time for a very new, post-apartheid state system. As 'crisis' though, its roots are also in what du Toit (1995: 351) identifies as the shared assumptions of those groups who challenged the apartheid state; populist participation and redistributive justice. That is, it is also as promises unfulfilled that this crisis takes form (P. du Toit 1995: 384-386, 417, 440), because it calls into question whether the 'new' South Africa is indeed the desired state and society (du Toit 1995: 439). Indeed it challenges newness itself. In such a situation, the state and the newly developing allied elites[42] need to find other ways of affirming both newness and its desirability.

The problem for this new state is that, while it has been able to achieve and make meaningful, individually and at a public level, the democratic and political form of nation, the successful completion of what *was* a modernist project, it now confronts an essentially postmodern society (Thornton 1996).[43] In turn, and again part of the problem, is that this is a society which actually *became* postmodern, not as some outcome of the development of capitalism to the late-phase being experienced and reproduced in many parts of the world, that is on the *successes* of modernism (cf Huyssen 1995). Rather, it became *post*-modern simply on the basis of modernism's inability to complete its project; its inability to bring to its peak the nation-state and the modernisation of South Africa. Such a society is now facing all of the crises of legitimation of the modern nation-state, but from a somewhat different starting point from say, the United States, Australia or the European countries. This provides a special role for transforming and newly developing elites of all kinds (P. du Toit 1995: Chapter 8).

Nation again emerges as a symbol and one discourse in which the state has sought to overcome these very particular contradictions and perceptions of crisis. In the face of tensions around the delivery of full redistributive justice in the name of citizenship, the state and its elites have turned to *other* narratives and symbols of nation. Citizenship *does* constitute the basis of a "we-language" (Hage 1993), but the history of fragmentation means that this we can so easily disintegrate under the very conditions that the postcolonial government inherited. In other words, the postcolonial condition contains both the impetus to unite and the seeds of disunity to undermine that impetus (Farred 1992; Werbner 1998). The state is committed to nation building, but in a postcolonial framework of unity in diversity which includes a concept of equality, and a *reality* of diversity and actual inequality of life-chances.

So what extra-political (and extra-economic) framework or set of strategies is available in post-apartheid South Africa to create other versions of shared nation, belonging and so engage all of the population? The very fragmentation and ethnic history of this postmodern society makes any notion of common culture, even the kind implicit in many liberal pluralist versions of multiculturalism. Culture itself remains a terrain of political struggle (Farred 1992: 219), so that any vision of nation has to include positively valued cultural difference. Then the *transcendence* of difference must be part of that nation's imagined content in order to construct the inclusive national collectivity of 'the people' - the nation's unity.

---

42 For discussion of the relationship between state and elites in nation building, see Orkin 1995; cf LiPuma 1997; P. du Toit 1995: 420; Chatterjee 1996: 224, 14.

43 Unlike the argument here, Thornton counterposes postmodern to postcolonial (cf Werbner 1996: 12), a formulation which minimises the role of the state in South Africa post-apartheid.

Temporality has become a crucial ground on which to think about these issues in South Africa and to imagine a South African nation; on it the political narratives of nation can also move into the complex cultural domain of other kinds of narratives (Bhabha 1990: 2). This is part of the significance of the appellation 'new' even where the content of that newness remains very unclear, certainly contested. Though the state is the agent of the future (Jewsiewicki and Mudimbe 1995: 6), it is built on and fundamentally engages, the past. The past is always important in national imagining. LiPuma (1997: 216, 217) argues for example that history is a privileged site in nation building, that is for imagining nation. He asserts that the postcolonial state must use details of the past (especially past violations) to present itself as the modern state which is also the "rescuer from colonial domination" (1997: 236). This would appear even more necessary in South Africa where the negotiated end to apartheid makes the 'rescue' unclear, because of the absence of a clear victory for the struggle. But what is important in this deployment of the past in postcolonial society's imagining of nation on the ground of the temporal is rupture; the imagining of the nation through decolonisation, the definitive break.[44]

In South Africa post-1994, the past was constructed as the nation's Other, that against which the present and future can be known and symbolically defined to construct a collectivity, the people, in the present as difference from the past. Implicated so strongly in imaginings and valorisations of nation, the past has become a political field, a site of struggle and contestation among agents and institutions as, first the state and various elites like the media and then other agents seek to control and shape it (cf Jewsiewicki and Mudimbe 1995: 8). Here arises the preoccupation with memory and narrative, particularly given the postcolonial character of South Africa.[45] This connection between nation building and temporality, particularly processes of history and memory, is not new. Jewsiewicki and Mudimbe (1995) for example, compare postcolonial Africa and post-Soviet Europe within a global frame to trace some of the state politics of managing the past in the name of the nation. "The twentieth century state presents itself as the unique political subject, the savior of the nation, legitimizing itself through the rationalization of the past, as produced by an invented historical discourse", they argue (1995: 5). Werbner et al (1998) turn to postcolonial Africa to study the complex relationship between memory and state nation building projects in a range of societies, each of which demonstrate what they term a "postcolonial memory crisis", over memorialising, forgetting but also "the very means and modes of remembrance" (1998: 1). LiPuma (1997) explores the efforts of various elites to construct a national history which precedes the multiple histories of this nation of diverse peoples. For them, a history which discovers a "primordial national identity" (LiPuma 1997: 217) integrates the many local histories of this heterogeneous society as a unity. In another part of the global family, Wood (1991), like many others, considers the role of the Holocaust (remembering, forgetting and narrativising it) in the construction of Germany as a nation since 1945. Lowenthal (1985) further considers the work of a whole range of representational institutions, from heritage bodies to national galleries, in constructing and contesting the national past (cf Adedze 1995; Watson 1995). Papadakis (1998) and others have considered how in Cyprus, individual and

---

[44] Werbner (1998: 75) contrasts the use of the past in old and young nations and argues that, "In Africa's nation-states, virtually all of which are young, the political origin myths usually imagine the founding of the nation in decolonisation."

[45] For more extended discussion of the relationship between memory, narrative and nation see Huyssen 1995; Geyer 1996; Jameson 1992; Santner in Wood 1991; Werbner 1997.

local histories get linked to the agendas and "abstract formulations" of various institutions and political parties to provide nationalist narratives (and the potential contestation around such processes). And of course, different narratives of the past which emerge in these and other situations can also often be exactly what threatens to divide a nation, tear it apart in a politics of difference (LiPuma 1997); in a counter-move to the suppression of some histories entailed in the construction of unifying narratives of nation.[46]

Media has often emerged as a central in the study of these complex interactions and processes. Anderson's (1992) is the formative work here, influencing much of the work in this field (see for example Chatterjee 1996: 216). But it would also seem clear that the emergence of electronic media has greatly extended the conditions and possibilities noted by Anderson. South Africa demonstrates that television and radio not only offer different possibilities for imagining nation, but their immediacy, accessibility and range also offer inclusiveness beyond illiteracy and imagined unity on an unprecedented scale (cf Barker 1999).[47] In turn, their rendering through the various genres of electronic media transforms the representation of the past in nation building processes.

Schlesinger (1991: 299) argues that today the modern state has to be understood as a "communicative space", making the media a political field in which cultural norms are imposed and resisted, where the state's interests are pursued but maybe also contested. Ma's study of the production of *Hong Kong Legend* takes into account the operation of the media field such that program makers are also influenced by economic and political relations to shape the new historical narratives in ways which accord with the perceived interests of capital and the Chinese state (1998). Although this is not its theoretical position, the study clearly demonstrates the usefulness of Schlesinger's communication space of the state. The media have become a crucial ground for nation building, not just in drawing members of the society into one polity as citizens, but also for creating connections between people despite diversity. This means that influential media personnel become numbered in those significant elite groups which have a role in opinion making and identity shaping.

Earlier we saw how, in South Africa, mainstream media like the SABC, have taken on board the state's nation building project and media have become the ground on which we find some of the strongest statements of nation as desirable and as a unity constituted out of diversity (Barker 1999: Introduction). In South Africa, a liberal democratic pluralist version of multiculturalism is apparent in the making of the same advertisements in different languages and the combination of at least two languages in continuity announcements on SABC television. Roome (1999) also discusses a highly popular situation comedy, *Suburban Bliss*, featuring two families, one black and one white, who after apartheid both work together and are neighbours. Again we have, here in comedy form, a quite intentional interrogation of the new South Africa, where 'new' stands for inclusion.[48] But the critiques of such programs

[46] For other examples of this tension see Brown 1996; Tachibana 1998; Healy 1997.

[47] Of course they also offer the potential to disturb identities, through the showing of alternatives and other lives on a *global* scale (cf Barker 1999; Thompson 1995: Chapter 7).

[48] Of the program, SABC publicity stated:

> In brief, it is the successful People Like Us concept moved into the new South Africa and taken one state further, as it explores the relations between two families, one black and one white, who not only own a business together, but also live next door to each other. As the series develops, it explores how the two families react to each others cultures, and focuses on their hopes and aspirations as the bonds of friendship between them grow stronger (*SABC Radio and TV*, No. 8, July/September 1993: 36).

that Roome also discusses indicate that good willed representation of this desired newness is probably not enough. Media have also powerfully picked up the theme of constructing, representing and exploring the past as the new South Africa's Other, that which helps define what the nation is today. The most obvious example is the media coverage of the Truth and Reconciliation Commission (TRC), in which the state and media came together to provide other agendas and other forms of imagining nation that did not rely entirely on redistributive justice and citizenship. Here media responded to the postcolonial imperative, constructing the past in particular ways through a construction and heightening of the processes of memory and narrative.

In the TRC, we find South African media providing a ground for the construction of and contestation over the past in the context of present state agendas and nation building processes. This connection between nation and the past in the TRC is explicit, as is the role of the media in the process. For example, in one magazine account we find said of the TRC, "Yet what makes the story worth re-telling is not only its heartrending elements - which make it a great story - but the fact that in it lies the seeds of a nation struggling to be born; a nation coming to terms with its past" (Mona 1998: 38). In the next two chapters I argue that the media coverage of the TRC can be understood as a form of nation building which goes further than the political form of nation so far achieved though proving problematic in ways discussed in this chapter. At the same time, it is predicated upon this form, particularly its values of equality and freedom, cast within a postcolonial frame. In South Africa the TRC is a primary example of what Werbner calls "politicised memory" (1998: 15); through memory work it locates the nation in the rupture with the past that he and his fellow authors see as common in African postcolonial states. Yet the above quotation, although it marks rupture, also suggests the need for *connection* with the past, here in the forms of interrogation, understanding and re-evaluation to overcome the problems of differentiation and fragmentation which make memory divisive of nation (cf Healy 1997). The mediated TRC can be understood as a ground on which individual and collective memories were articulated, elaborated, harnesssed and framed in an attempt to forge a new vision of 'the people' out of the apartheid past by constructing a representation of *history* which can be shared, not in the ethnic sense referred to by Hobsbawm (1990), but in the sense that it encompasses all, regardless of the experience of apartheid. Like citizenship itself, such a history has the potential to locate and transcend the multiple histories, narratives and identities of a fragmented postmodern society and so provide the ground of nation as imagining Oneness (LiPuma 1997). In its conception, it seeks to relocate people in the *same* time and space, both in the past and, importantly, for the future. In it narrative, the multiple "'nows' lived alongside each other" through which Pechey (1994: 155) characterises the fragmentation of South Africa, are brought into a structured relationship with each other.

In: Advances in Communications and Media Research, Vol. 3 ISBN 1-60021-188-7
Editor: Patula E. Grossmark, pp. 61-70

Chapter 4

NEWSPAPER EMPLOYEE MANAGEMENT IN A DIVERSE POPULATION

Meta G. Carstarphen
Gaylord College of Journalism and Mass Communication
University of Oklahoma, Norman, Oklahoma, USA

TRADITIONS AND INFLUENCES

From its earliest origins, the newspaper presence in the Americas has been adventurous, spirited and, as a matter of history, operating in the midst of often untapped diversity. When British printer Benjamin Harris published his one and only issue of *Publick Occurrences, Both Foreign and Domestick* in 1690, he launched the first newspaper among the English colonies along what is today the east coast of the United States. Sloan and Williams (1994) noted that Harris' journalistic work stood out for its "emphasis on political and dissident Protestant works" (p. 3), reflecting some of the regional politics and British dominance of the emerging states.

But Harris' effort, while historic, was not the first recorded work of journalism outside of European borders after the mid-15th century evolution of the Gutenberg Press. Other scholars note that the first printing press arriving in the Americas landed in Mexico City more than a century earlier, perhaps in 1535. This press was established by Spanish printer Juan Cromberger of Seville (Wilson, Gutiérrez and Chao, 2003). In 1541, that same press issued what may have been the first news report in the Americas, when writer Juan Rodriguez issued a missive giving an account of natural disasters that had devastated a city south of Mexico City. Written in Spanish and informed by the Catholic Church's influence, this news report combined vivid detail with moral interpretation and reflected the language and probably the concerns of area residents.

What these historic examples show, be they from Europe or the Americas, is that whoever owns the medium, controls the content. Now, as then, managing the news organization has direct implications upon the creation of content. And, as will be demonstrated below, the "business" of newsgathering in today's world has never been more important, nor more contested.

In the media of the 21st century, both who works in the media and even how the news is presented, may depend much upon whoever owns the media. Today, large local and national newspapers struggle with finding ways to attract, employ and retain diverse staffs because, as will be shown below, time and time again history has shown that it matters. Increasingly, newspaper managers must balance expending resources toward this imperative against the challenges of meeting aggressive competition from an ever-expanding media landscape, increased production costs, and stagnant--even falling--readership rates.

Readership stagnation is not restricted to the United States alone. Based upon a survey of newspapers in 208 countries by the World Association of Newspapers (WAN), global newspaper circulation showed a 0.12 percent decline in 2003, compared to 2002. According to the 2004 study, the U.S. and Japan were among the countries showing flat or declining circulation figures, in contrast to growing circulation in places such as China and India (Bhatnagar, 2004).

Still, despite these competitive challenges, or perhaps because of them, U. S. newspapers know that their future depends upon how well they become vital to communities that are increasingly becoming more ethnically diverse. According to a December 2003 announcement, the census bureau projected that U.S. population growth had exceeded 290 million people (U. S. Census, 2003b). In an earlier analysis, the U.S. Census Bureau completed a county-by-county population estimate, showing for the first time where ethnic and racial shifts were occurring on local levels (U. S. Census, 2003a). The net results were that the demographic assumptions about the communities they served, which had sustained newspaper management decisions for years, were not necessarily adequate for present challenges, especially the economic ones. And, as will be shown below, the demographics of newspaper employees did not reflect the communities served by these newspapers. As one scholar noted in a report about diversity and the newspaper business, *employee* diversity was one of the keys to improving newspaper revenues among burgeoning, new communities of readers because of the "credibility" and "fresh perspectives" they could bring to bear (Sullivan, 1999, 7).

During the last century, important touchstones helped define the current urgency and have helped to pave the way toward possible solutions.

Commissions and Criticisms

Since its origins, the United States and its media have struggled to come to terms with diversity and its implications. What role have newspaper journalists played in exacerbating that problem through their work in the mass media? This role has been received with either benign neglect or intense debate.

Hutchins Commission

The first significant and broad analysis of newspaper performance occurred before mid-century. In 1947, a council of 13 intellectual and business leaders, headed by Robert M. Hutchins and funded by Henry R. Luce, studied the role of "a free and responsible press" in

American society.[1] While considering far-ranging questions related to the content, regulation and governance of the press, these men wrote, albeit in oblique language, about issues that anticipated today's concerns about diversity and fairness. Speaking as if for the larger society, commission members presented "five ideal demands" of the press. In these demands, they challenged the media to be accountable to their public in specific ways (Gross, 1966). Among these five demands were two imperatives that elevated concerns about representation and the media:

(1) give "a truthful, comprehensive, and intelligent account of the day's events in a context which gives them meaning," and
(2) project "a representative picture of the constituent groups in the society" (p. 30).

Ironically, while this group ostensibly upheld an ideal of representative journalism to the newspaper industry, a standard practice of exclusion and invisibility operated among the press organizations it represented. The majority press for decades had followed an unstated policy of omitting coverage of minorities, especially African Americans, from its news pages, an industry-wide practice of absence which Martindale (1986) noted and which transcended prejudices within individual reporters. Furthermore, whenever African Americans did appear in the news, most noticeably as the Civil Rights movements of the 1950s and 1960s grew, their stories were told as "discrete events" and as "controversy" (Martindale, p. 32). These frameworks of event-driven and conflict-defined news are familiar constructs within traditional "news values"; however, when they represent the primary ways in which the news of people of color are reported, they communicate a distorted reality about these communities to the readership at large.

In the 20th century, what propelled the increased coverage of minorities into the news reports of major media were the dramatic and contentious struggles of African Americans during the Civil Rights era. This period during the 1950s and throughout the 1960s marked a discernible increase in coverage of African Americans in major newspapers across the country (Keever et al, p. 87). Thus, Civil Rights issues emerged as a societal force, wedged between the Hutchins report and an even more dramatic study that would emerge in the late 1960s. Nearly two decades after the 1949 Hutchins report, a new context formed for the next watershed study of media's role and impact. These social issues shaped the content and focus of the *Report of the National Advisory Commission on Civil Disorders,* or Kerner Commission Report (The Kerner Report, 1988).

Kerner Report

President Lyndon B. Johnson convened this 1967 panel at the height of national distress about urban-centered riots that had been marked by a series of explosive confrontations between whites and blacks. As the panel examined the relationship of the mass media and the violence, members concluded that their role was both profound and disturbing. The

[1] Hutchins was President of the University of Chicago and Luce was publisher & editor-in chief of *Time* magazine. For an in-depth discussion of the report and its participants, see "Realigning Journalism with Democracy: The Hutchins Commission, Its Times, and Ours," by Annenberg Senior Fellow Stephen Bates.

Commission also blasted the press for its singular, exclusionary vision: "The media report and write from the standpoint of a white man's world . . . .a press that repeatedly, if unconsciously, reflects the biases, the paternalism, and the indifference of white America" (p. 366). This same report warned against the imminent fracture of the American society into "two societies, one black, one white" (p. 1). The Commission reported a dichotomy that, unfortunately, reduced the complexity of diversity into stark and opposing symbols that arguably have influenced journalism's race reporting ever since.

Newspapers Face the Diversity Challenge

ASNE Year 2000: Ten years after the Kerner report, the American Society of Newspaper Editors launched what would become an annual census of employment within its ranks, using these survey results to highlight the industry's challenges and to prod members into more aggressive activities of inclusion. In 1978, out of a total workforce of 43,000, ASNE (2004) recorded that 1,700, or 3.95 percent, were minorities. With those dismal numbers before it, and with what seemed like the generous span of time ahead of it until the next century, ASNE established the "Year 2000" goal. Widely reported, this effort doggedly presented a challenge to "newspapers to achieve diversity in their newsrooms equivalent to the U.S. minority population by the year 2000 or sooner" (Jennings, 1995, p. 22).

By 1995, ASNE recorded slow, incremental progress in the total numbers of journalists of color, as well as patterns that described some parameters of that progress. For instance, as Hernandez (1996) noted, over half, or 52 percent, of the minority workforce were reporters, followed by 18 percent who were supervisors, 16 percent, copy/layout editors, and 13 percent photographers. In the same report, larger newspapers, with circulations of 100,000 or more, were shown to have the best success: 68 percent of all minorities were there, compared to 38 percent who worked for papers with circulations under 100,000.

Finally, a very dramatic difference between markets showed up in the involvement of students of color in internships. The ASNE survey revealed that at newspapers under 5,000 circulation, students of color made up only 8.3 percent of the interns, compared to larger newspapers with circulations of 500,000, where minorities occupied 55.1 percent of available internships. As invaluable pipelines into the profession, the paucity of minorities in internships at smaller papers has had an impact on the diversity of the permanent workforce at these papers, as well as -- more indirectly -- the opportunities for students of color to gain experience in the field.

In its 1998 census, ASNE (2004) showed that although the percentage of journalists of color in the professional workforce at papers with 5,000 circulation or less had doubled since 1978, the current 6.6 percentage reflected slim representation (p. 22). Still compared to 1978, or certainly 1968, the profession by 1995 was experiencing more diversity than ever. But, was it reflecting more diversity in its news pages?

News Coverage

Increasingly, scholars, professional observers and audiences themselves were less than enchanted with the news they were reading, receiving and viewing. For example, in one study following the 1992 Los Angeles riots (Carstarphen and Kim, 1998), African American and Korean American businessmen in another major city found that newspaper and other media coverage contributed to their *own* biased opinions about their respective communities. In a scenario eerily reminiscent of the 1967 disturbances, major media reports were seen as framing negative portrayals of both minorities, but especially those of African Americans (p. 228).

Entman's (1992) analysis of local Chicago newscasts documented a prevalence of crime and political reporting about African Americans, which in his view, served to fuel "modern racism" (p. 343). According to Entman, modern racism was a phenomenon characterized by hostility and rejection directed toward minorities. In other words, as white audiences saw repeated stories about blacks in trouble or involved in political protests, they were able to generalize about behaviors they believed to be typical of African Americans. An exception was when Black anchors delivered the same news stories showing other, less successful Blacks involved in crime or protests. In those cases, the subtle message that some blacks could overcome obstacles unintentionally reinforced a perception that racial bias did not exist at all (p. 358).

In a later work where television reporting of violence was compared to such coverage in newspapers (Entman and Rojecki, 2000), results showed that racial stereotyping in newspapers appeared less blatant. Some reasons attributed to the differences included fewer visual illustrations and a target audience whose more affluent backgrounds dictated less sensationalism in their news offerings.

The concerns about how the news business could improve its reporting of diverse communities were shared broadly across media platforms. One evidence of this shared concern among the growing numbers of reporters of color came when four professional groups joined forces in the 1990s to place a different spotlight on diversity.

Unity 94

In July 1994, the historic Unity '94 conference met, convening together for the first time members of the Asian American Journalists Association, the National Association of Black Journalists, the National Association of Hispanic Journalists and the Native American Journalists Association. The 56-page "News Watch" report unveiled at that gathering criticized mainstream media's coverage of people of color as being "riddled with old stereotypes, offensive terminology, biased reporting and a myopic interpretation of American society" (Stein, 1994, pp. 12-13).

As reporters of color continued to voice their concerns about journalism, they promoted a broader interpretation of diversity and the news. Though employing more minorities continued to be important to them, journalists of color spoke with the conviction and frustration of "insiders" who nonetheless still felt thwarted by the elusive goals of parity and fairness in the news. Among the reporters' concerns was their disappointment at how the media tended to view diversity as a just a "black and white" issue (Terry, 1994, p. 26) and their awareness that their own ethnic communities were pained on a regular basis by the work emanating from the media. The same report included a 1994 USA/CNN/Gallup poll that showed 62 percent of the African Americans queried reported being "upset" at least once a week by news coverage, with 39 percent of Latinos and 29 percent of Asians concurring. When asked to assess the impact of local newspaper coverage on race relations, 47 percent of the African-American respondents felt it "hurt" more than it helped, with 24 percent of Latinos, 19 percent of Anglos and 12 percent of Asians in agreement (p. 27).

While these reactions may seem like a curious indictment of media diversity from the very public it is designed to serve, they are instead a pointed reminder that serious flaws in news content run deeper than the cosmetic presentation of the news.

Professional Issues, More Challenges

In its "1998 Newsroom Census" on diversity (against a national population of 26 percent minority), ASNE reported that, of a total workforce of 54,700, journalists of color represented 11.46 percent of newsroom employees, noting a small increase from 1997's 11.35 percentage. (Jennings, 1995, p. 23). With the new century looming, ASNE made a dramatic, and strategic, shift. While acknowledging that the industry would fall far short of its year-2000 goal, ASNE proposed an extension of its diversity timetable and a broadening of its intent. The new parity would seek a 20 percent minority representation by year 2010, and its figures would include women, gays, lesbians and the disabled within its target population (Editors' Group, 1998, p. 54). This dramatic shift in focus reflects the continuing difficulties the profession has in maintaining diversity and incorporating it within its own ranks, while reporting on an increasingly vocal and pluralistic society.

Similarly, the Radio-Television News Directors Association (RTNDA) reported that in 1997, the total minority workforce in local television news increased from 19 to 21 percent from the previous year. During the same period, however, the percentage of minority news directors working in both radio and television dropped from 9 to 8 percent (Flander, 1997, p. 27). Additionally, the report found that when there was a minority news director, he or she was eight times more likely to work at an independent station than at a network affiliate (p. 27).

Not only was the presence of minority "managers" decreasing, but the barriers to their working for a major network or its affiliate were tremendous, as they were eight times less likely to be hired at such stations. In fact, in quoted anecdotes, some candidates related specific instances where they felt being a minority "closed doors" on certain opportunities within their careers (p. 27).

While professional organizations continue to push for increased diversity in the journalism workforce and the management of these organizations, there exists, nevertheless, a dissonance. These differences exist (1) between the expectation of how much diversity there should be and the reality and (2) between the daily experiences of most Americans in their communities and the reports coming back to them from their media. The latest industry surveys continue to show modest growth of people of color in the media workforce, but still reflect considerable barriers to these minorities' having positions of influence within mainstream media. Backlash and resistance to change certainly contribute to this crisis. One author (McGowan, 2001) took diversity and newspaper content to task, blaming its failings on a wide range of issues. These issues included having newsroom staffs being held hostage by diversity goals and, by extension, inferior content. And the widely publicized firing of Jayson Blair from the *New York Times* fueled more debate and controversy over how much the role of pressure to have a diverse staff contributed to this issue (Loop, 2004).

All of these strategies lend subtext to these unresolved discussions in journalism newsrooms and their staffs. Yet, supporters and critics alike of newspaper diversity continue to keep the issue alive in the marketplace of ideas. Even so, in its 2003 employment census at U. S. daily newspapers, ASNE reported that figures showed a modest half of one percent increase in diversity over the previous years. Parity with the nation's demographic profile is still a distant goal, and the news organization noted that this growth of 12.94 percent still "lagged behind the 31.7 percent minorities in the U.S. population" (ASNE, 2004, 1).

Diverse Past, Diverse Futures

Still, efforts to propel newspapers into a diverse and, by extension, economically secure future persist. Fresh from her election to the ASNE presidency, experienced editor Karla G. Harshaw announced her intent to make connecting with communities of color her leadership focus (Harshaw, 2004). And, she is not alone. Industry-wide responses to the changes impinging upon the profession include initiatives affecting both internal operations and public outreach. Consider the following new or continuing initiatives:

- Individual newspaper groups, individual newspapers and some state-wide coalitions have diversity programs in place (Loop, 2004);
- "Project GOLD" of the Newspaper Association of America (NAA) offers newspapers research, training and specific strategies designed to leverage diverse content with diverse business opportunities; the NAA is a nonprofit organization representing the newspaper industry;
- The NAA also sponsors a Minority Fellowship program offering training for journalists of color who show managerial potential;
- The "Parity Project," launched in 2002 by the National Association for Hispanic Journalists (NAHJ), partners the affiliations and resources of this professional journalists' group with media in order to increase the employment of Latino journalists working in daily newspapers and English-language television stations;
- Ongoing programs by the National Association of Black Journalists (NABJ) Media Institute provide media professional development and technical training for students and professionals seeking career advancement;
- The Executive Leadership program sponsored by the Asian American Journalists Association (AAJA) offers training and support for mid-careerists wanting to advance from newsroom to administrative careers;
- A Tribal Media Institute to be launched by the Native America Journalists Association (NAJA) will help promote more career opportunities for Indian Journalists, as well as advocate for better coverage of Native American issues;

- Through the Dow Jones Newspaper Fund, nearly 40 summer workshops convene in partnership with area media to target minority high school students interested in journalism careers;
- Since 1991, the Chips Quinn Scholars program has placed more than 800 young journalists of color in daily newspaper newsroom internships, allowing them to access opportunities for journalism careers;
- Once a year, the Associate Press Managing Editors (APME) and ASNE team up to sponsor a National Time-Out for Diversity and Accuracy campaign, designed to heighten awareness within newsrooms about the impact of diversity within their communities;
- Journalism "think tanks," such as The Poynter Institute, The Maynard Institute, and the Freedom Forum, regularly provide programs highlighting diversity practices and encouraging newspaper staffs at all levels to apply proactive practices at their home organizations.

Ultimately, the near future projects a slow march toward more representative staffing, management and ownership within newspapers. One encouraging indicator from the 2003 ASNE census showed that one measure--minority internships at newspapers--may reflect parity. With the overall percentage of people of color in the U.S. estimated at 31.7 percent in 2003, ASNE found that the number of interns of color in daily newsrooms equaled 32.2 percent during that same year (Fitzgerald, 2004).

Employee diversity in newspapers will yield benefits to all readers and all media. Walker (2004) reported that circulation for Spanish-language newspapers has increased nearly four-fold over 13 years, while English-language daily newspapers have lost 11 percent of their circulation since 1990.

Attracting multi-ethnic readers to the mainstream publications is a priority. But, many mainstream newspapers aggressively pursue new audiences by creating niche editions of their publications, relying heavily upon the talents of journalists of color. One example includes the Spanish-language daily newspaper, *Al Dia,* which debuted in Dallas in October 2003. Longtime journalist and former NAHJ president Gilbert Bailon serves as *Director y Editor* for this Monday-through-Saturday newspaper, as well as its accompanying website, *Aldiatx.com.* Media corporation Belo Inc. created *Al Dia,* a distinctive complement to its English-language sister publications in the north Texas region, such as the *Dallas Morning News* and the *Denton Record-Chronicle.* Similarly, Knight-Ridder Corporation, owner of such newspaper properties as the *Detroit Free Press* and the *Philadelphia Inquirer*, owns Miami-based *El Nuevo Herald,* led by Executive Editor Humberto Castelló. Encouraged by the new technological possibilities offered online, Internet websites offer expanded opportunities for ethnic audience media, not fewer (Carstarphen and Lambiase, 1999), as a result promoting more diversity, not less.

As the second oldest mass medium, behind books, newspapers remain influential in communities across many borders. In a groundbreaking and comprehensive report about American media released in 2004 by the Project for Excellence in Journalism, researchers noted that ethnic media are experiencing rapid growth (Project for Excellence, 2004). As ASNE continues its annual census of newsroom employment and diversity, the progression toward true parity will continue to be documented, updated and celebrated as a necessary evolution of American media.

References

American Society of Newspaper Editors. (2004, April 20). Newsroom employment drops again; diversity gains. *News release* [Electronic version]. Retrieved May 13, 2004 from http://www.asne.org/index.cfm?id=5145

Bates, S. (1995) *Realigning journalism with democracy: The Hutchins commission, its times, and ours*. Washington, DC: The Annenberg Washington Program in Communications Policy Studies of Northwestern University. Retrieved March 7, 2004 from http://www.annenberg.nwu.edu/pubs/hutchins/

Bhatnagar, A. (2004, May 31). Newspapers circulation figures dip worldwide. *The economic times*, p. 279. Retrieved June 6, 2004 from http://economictimes.indiatimes.com/articleshow/msid-711169,curpg-2.cms

Carstarphen, M. G., and Lambiase, J.J. (1998) Domination and democracy in cyberspace: A report from the ethnic/gender margins. Bosah Ebo (Ed.), *The cyberspace ghetto*, (pp. 121-135). Westport: Greenwood Press.

Carstarphen, M. G., and Kim, T. G. (1998) Who's the victim? Intercultural media perceptions between African-American and Korean-American businesspeople in Dallas. Y. R. Kamalipour and T. Carilli (Eds.), *Cultural Diversity and the U.S. Media* (pp. 217-229). Westport: Greenwood Publishing Group.

Editors' group proposes change in minority hiring goals at newspapers. (1998, April 27). *JET Magazine*, 95, 54-55.

Entman, R. (1992). Blacks in the news: Television, modern racism and cultural change. *Journalism quarterly, 69*, 341-361.

Entman, R., and Rojecki, A. (2000). *The Black image in the white mind: Media and race in America*. Chicago: University of Chicago Press.

Fitzgerald, M. (2004, April 20). Minority journalist representation nudges up slightly. *Editor and Publisher* [Electronic version]. Retrieved May 20, 2004 from http://www.editorandpublisher.com/eandp/search/article_display.jsp?vnu_content_id=1000491572.

Flander, J. (1997, October,). Tales from the trenches. *Communicator*. Washington, D.C.: Radio-Television News Directors Association, 27.

Grangenois, M. (1993, April). Whites need not apply? *Quill, 81*, 23-24.

Gross, G. (Ed.). (1966). *The responsibility of the press*. New York: Simon and Schuster.

Harshaw, K. G. (2004, March). Connecting with communities and cultures. *The American editor, 79*, 2. Retrieved April 28, 2004 from http://www.asne.org/files/tae200403.pdf

Hernandez, W. (1996, May 25). Growing Ever So Slowly. *Editor and Publisher, 129*, 23.

Jennings, V. (1995, March). Minorities are now 10.9 percent of the newsroom work force in the U.S. *ASNE bulletin, 766*, 22-23.

Keever, B. A. D., Martindale, C., and Weston, M. A. (Eds.), (1997) *U.S. news coverage of racial minorities.* Westport: Greenwood Press.

Loop, T. (2004, February). The direction of diversity. *presstime, 26* [Electronic version]. 4823 words. Retrieved March 31, 2004 from LexisNexis print delivery.

Martindale, C. (1986). *The white press and Black America.* New York: Greenwood Press.

Martindale, C., and Dunlap, L. R. (1997) The African Americans. In B. A. D. Keever, C. Martindale and M. A. Weston (Eds.), *U.S. news coverage of racial minorities* (pp. 61-145), Westport, CN: Greenwood.

McGowan, W. (2001). *Coloring the news: how crusading for diversity has corrupted American journalism.* New York: Encounter Books.

Report on the National Advisory Commission on Civil Disorders: Special introduction by Tom Wicker. (1968). New York: Bantam.

Rowe, S. (1997, September). Looking through another lens is part of credibility project. *The American editor* [Electronic version]. Retrieved April 5, 2004 from the online archives for the American Editor http://www.asne.org/kiosk/editor/97.sept/rowe1.htm

Sloan, W. D., and Williams, J. H. (1994). *The early American press, 1690-1783.* Westport: Greenwood Press.

Stein, M. L. (1994, August 6). Racial stereotyping and the media. *Editor and Publisher, 127,* 12-13.

Sullivan, D. (1999). Linking diversity to business results. GOLD: Growth opportunities by leveraging diversity. The GOLD network: GOLD 'n' Press. Retrieved May 1, 2004 from http://www.naa.org/diversity/gold/press/Linking_Diversity_to%20Biz_Results.cfm

Terry, C. (1994, September). Different drummers. *presstime, 16,* 26-30.

U. S. Census, Bureau of. (2003a, September 18). First county population estimates by race and ethnicity since 2000 show diversity changing local demographics. *News release* [Electronic version]. Retrieved April 20, 2004 from www.census.gov/Press-Release/www/2003/cb03-143.html

U. S. Census, Bureau of. (2003b, December 18). U. S. population passes 290 million: Mountain and coastal states fastest-growing. *News release* [Electronic version]. Retrieved April 20, 2004 from www.census.gov/Press-Release/www/releases/archives/population/001624.html

Walker, J. (2004, April). Study pinpoints major trends shaping news media landscape. *O'Dwyer's PR services report, 18,* 41.

Wilson, C., Gutiérrez, F., and Chao, L. M. (2003). *Racism, sexism and the media: The rise of class communication in multicultural America* (3rd ed). Thousand Oaks: Sage.

In: Advances in Communications and Media Research, Vol. 3 ISBN 1-60021-188-7
Editor: Patula E. Grossmark, pp. 71-84

Chapter 5

REGULATION OF NEWSPAPER OWNERSHIP

Todd F. Simon
Kansas State University, Manhattan, Kansas, USA

There is no official regulation of newspaper ownership in the United States, at either the federal or state level. Antitrust law, both federal and state, provides a basis for preventing changes in newspaper ownership on a case-by-case basis, but there is no systematic oversight about who or which companies own which newspapers in what markets.

The only two direct laws or policies about newspaper ownership are exceptions to the norm. The Newspaper Preservation Act allows an exception to the antitrust laws (Newspaper Preservation Act, 1970). It allows the owners of competing dailies in the same city to combine business operations, thus creating a monopoly, which would otherwise violate antitrust law. The Federal Communications Commission has prohibited combined ownership of a daily newspaper and a commercial television station in the same market for decades. In 2003 the FCC sought to revise this cross-ownership rule, eliminating it in most cities, but after extensive hearings, a challenge in court, and a partial reversal by a Federal appeals court, it abandoned the proposed changes in early 2005.

Antitrust law has proven to be an almost toothless way of regulating newspaper markets. The deregulatory zeal of the past 25 years has meant that very few ownership changes have been challenged by the U.S. Department of Justice's Antitrust Division. And those few are not typical of transactions where newspaper ownership changes. There is usually evidence of a direct intent to shut down newspapers or to otherwise adversely affect consumers when the division steps in.

ANTITRUST LAWS

There are two main U.S. antitrust laws, the Sherman Act and the Clayton Act, but only the Clayton Act is designed to address ownership acquisitions and changes. The Sherman Act focuses on attempts by existing owners to monopolize a market, either alone or by working with others (Sherman Act, 1890). It has largely been limited to such abuses as price fixing or attempts to exclude others from a market, and so has seldom been used as the basis for action against ownership changes.

Clayton Act

The Clayton Act specifically aims at mergers, purchases, and consolidations, and allows a challenge when an acquisition serves "substantially to lessen competition or tend to create a monopoly" (Clayton Act, 1914). Although the language of the statute adopted by Congress plainly anticipated a broad interpretation, the actual operation of the statute was largely left to the DOJ Antitrust Division and to the Federal Trade Commission (FTC) to determine. As is true with most administrative regulatory agencies, the strictness of the interpretation varies with presidential administrations and with the political views of the times. Since the Carter administration from 1977 to 1981, the sense has been that competition in the marketplace is preferred to government intervention in the behavior of firms. As a result, the Federal government has seldom objected to acquisitions of newspapers absent clear evidence of economic harm to competitors and/or consumers.

The Clayton Act, however, also allows private actions. These can either be brought by competitors or by consumers. In general, though, these actions have gone nowhere. The explanation for the dearth of government actions and for the failure of private actions lies in how the Clayton Act has been applied. Anyone challenging a merger must show that competition is lessened in both the relevant "geographic market" and the relevant "product market." Courts have determined that daily newspapers are an entirely separate kind of product from paid circulation weekly newspapers, free-distribution newspapers, and other publications (Lacy and Simon, 1993). The analysis almost guarantees that no cases will succeed.

Product Market

Dividing newspapers into distinct product categories may be historically accurate, but in today's media environment it is astonishingly naïve. It assumes no competitive overlap between types of media in a market or between types of newspapers in a market. But it is a given that all local media compete for the same advertising dollars, for the attention of the same consumers, and, if they have news and public affairs units, to provide reporting and commentary within the market (Beaudoin and Thorson, 2002). Yet case after case follows precedents, some half a century old, which created the separate product analysis (U.S. v. Times-Mirror, 1967; Community Publishers, 1995). In that case, a court decided that the product market for daily newspapers in the metro Los Angeles area would not include daily newspapers in the suburbs. That did not make sense in the 1960s—the papers were trying to serve the same people. The product market paradigm is predicated on the idea that newspaper products, plus other media products, are not substitutable. To the extent that each is distinctive, the argument has some merit. But the analysis simply ignores the ways that newspapers and media products are similar. The number of communities in the U.S. that fit the traditional product market analysis—a single daily or weekly newspaper plus a non-duplicative television station and radio stations—cannot be large. And the communities themselves must be tiny.

A recent acquisition review by the Antitrust Division shows that the narrow definition of product market is alive and well. The New York Times Co., which owns the daily Boston Globe, recently purchased the free-circulation daily tabloid Metro Boston. The Globe has circulation of more than 451,000, the Metro 180,000. Under the agreement, the Globe will provide some news content to the Metro. The two papers will also begin cross-promotion and joint sales. When the proposed purchase was announced, the publisher of the daily Boston Herald complained, as did some consumers, triggering review by the Antitrust Division. Despite "investigation" by the DOJ, the deal was approved in less than two months (Antitrust investigations, 2005; Jurkowitz, 2005; With DOJ's antitrust OK, 2005). Apparently free-circulation and paid-circulation daily newspapers are sufficiently dissimilar to qualify them as different types of products; by this analysis, a low-priced subcompact car and a luxury Mercedes should not be considered as within the same product market. Localism concerns raised by consumers seemed thin here: publicly traded Metro International, owner of Metro Boston, publishes 54 newspapers on four continents.

Geographic Market

The analysis of the geographic market is disconcerting too. Circulation areas are considered to be distinct, sharply defined, and strictly local. It assumes either that only one newspaper product circulates within a geographic market, or that a slight amount of overlap in the geographic market does not create an anti-competitive effect. Such an analysis runs counter to what research has told us for decades (Bridges, Litman and Bridges, 2002). As a result, acquisitions of newspapers–even by other newspapers operating in the same community or metro area–is not seen as a problem (Lacy and Simon, 1997).

The latest confirmation that the notion of geographic market is no meaningful impediment to mergers was Gannett's acquisition of Hometown Communications in early 2005. Hometown owned 62 weekly or twice-weekly newspapers, 24 community phone directories, one daily newspaper, and a variety of specialty publications. Hometown operated in Michigan, Ohio, and Kentucky. Gannett is the largest newspaper publisher in the country in terms of number of papers, with 102 dailies and 800 other publications. Among its papers are the *Detroit News*, the *Lansing State Journal*, and the *Cincinnati Enquirer*. The *News* and *Enquirer* are published under joint operating agreements and are considered the dominant papers in those markets. The *State Journal* is the only daily in Lansing, Michigan. Hometown's publications primarily serve the Detroit metro area, the Cincinnati metro area, and the Lansing metro area. When the planned purchase was announced, the DOJ said it would give it a close look (Ives, 2005), apparently because it would leave those markets dominated by a single owner. But the deal was approved in a short time (Gannett, 2005).

Another case taking a narrow view of media competition in local markets was *Berlyn v. Gazette Newspapers*. Gazette was owned by the Washington Post Company. Since 1992 it had acquired 31 suburban weekly or semi-weekly newspapers in the Washington, D.C. metropolitan area. A 10-paper sale in 1999 triggered a lawsuit by Berlyn, which published two newspapers in the Maryland suburbs. The case featured claims of violation of both the Sherman Act and the Clayton Act. What is most notable, though, is how the court analyzed the geographic market. The trial judge doubted that the plaintiff's defined market—three counties adjoining the District of Columbia—could actually be called a suburban newspaper market. One county remained largely rural, while another was slightly rural (*Berlyn*, 2002). Berlyn also claimed price discrimination under the Clayton Act, in this case charging different advertising rates solely to disadvantage Berlyn. While the judge apparently agreed that there may have been some dicey pricing, the claim went nowhere; the price discrimination provisions only apply to the sale of commodities, and newspaper advertising had long been deemed a noncommodity by the courts.

The *Berlyn* case also claimed that Gazette had conspired to reduce competition by being a partner in a suburban D.C. advertising sales agency, Press Network. That claim failed because the half-purchase of the agency was not the basis for the case; rather, a recent purchase of other newspapers had been. There was no showing that Press Network itself engaged in any activities that reduced competition. The case had an additional ironic twist. The judge sealed much of the record at the request of the defendants, who claimed it contained confidential business information (Sullivan, 2002).

The biggest newspaper company merger in recent years was approved in February 2005. Lee Enterprises agreed to buy Pulitzer Inc., with the sale to be completed in June 2005. Lee owns 44 daily newspapers plus dozens of weekly, specialty, and shopper-type publications. Pulitzer owns 14 daily newspapers and 100 non-daily publications, including 38 metro-area non-dailies in the St. Louis metro area. The deal makes Lee the fourth-largest newspaper chain or group in the country. Lee was competing with Gannett to buy Pulitzer. The sale agreement was announced in January 2005 (Fitzgerald, 2005); DOJ approval came the next month (Pulitzer Deal, 2005). Although the merger features communities with multiple publications in an overlapping geographic market, there was no claim, as in the *Washington Post* and *Boston Globe* purchases, of anticompetitive effect. Perhaps there were no newspaper competitors left in the market to complain. Two Pulitzer stockholders threatened to file suit, but on the ground that the stock price was too low, not on antitrust grounds. One effect of the merger is that now almost all of the daily newspapers in the state of Wisconsin are owned either by Lee, Gannett, or Journal Communications (Knutsen, 2005).

One may wonder why mega-deals such as the Lee-Pulitzer merger get approved so quickly. When the size of a sales transaction exceeds $50 million, the Hart-Scott-Rodino Antitrust Improvements Act requires that notice and disclosure be made to the Department of Justice's Antitrust Division (Hart, 1976). The intent of the statute was to slow merger activity by requiring filing and disclosure, thus allowing the DOJ a chance to look things over closely before the union is consummated. If satisfied there are no antitrust problems, the DOJ signs off on the merger. If there are doubts, DOJ can squelch the deal or require that it be structured differently. At least where media mergers are concerned, though, Hart-Scott-Rodino has worked as a method of fast-track approval rather than a brake (Klein, 1997; Overture-Yahoo, 2003). The review period is 30 days. If DOJ says nothing and the period expires, an acquisition may proceed. A silent approval by DOJ undoubtedly deters private lawsuits as well.

Compared to corporate merger activity overall, acquisitions in the newspaper industry seem fairly modest (Myerson, 2005). But the effect of newspaper mergers is of a different kind than are most mergers. When Phillips 66, Conoco, and Union 76 became a single gas and oil company, the antitrust concern is primarily to assure that prices and supplies are not manipulated. But with newspapers, consumers have other concerns. Newspapers cost them little to buy. Many are free. But consumers wonder what effect distant, conglomerate ownership will have on the news and public affairs product they pick up at the store or have delivered at home (Chambers, 2003).

One significant antitrust action was pursued by the Department of Justice in recent years. Village Voice Media and NT Media, two corporations that published many of America's largest free-circulation alternative metro-area weeklies, agreed in 2002 to swap markets. The two companies competed head-to-head in Cleveland and Los Angeles. Village Voice's *LA Weekly* was the largest alternative paper in the country, with circulation of 215,000. NT's *New Times LA* had a circulation of 120,000. In Cleveland, Village Voice's *Cleveland Free Times,* with 80,000 circulation, led NT's *Cleveland Scene.* The competition in both markets drove advertising rates lower and newsroom investment higher. To achieve monopoly status, the two companies agreed that Village Voice would leave the Cleveland market to NT and NT would leave the Los Angeles market to Village Voice.

Sherman Act

Generally, an agreement to allocate markets to achieve monopoly status is considered a per se, on its face, violation of the Sherman Act. A federal trial judge agreed in U.S. v. Village Voice Media in 2003. What is most striking here is that the two defendants entered into a formal, written contract to divvy up the two markets. Village Voice also paid NT $9 million cash. These activities suggest that the two companies must have believed that antitrust actions against newspapers would not ever be pursued again (U.S. v. Village Voice, 2003). In the past, evidence of collusion was enough for two or more newspaper defendants to lose an antitrust case (Lacy and Simon, 1993), even if there was not definitive documentary proof.

In the 2003 case, the remedy imposed by the judge in a proposed final judgment without a trial essentially reinstated the status quo. Even though Village Voice's Cleveland paper had been closed and NT' Los Angeles paper also closed, the firms were ordered to start them up again. Assets transferred between the two were ordered sent back. The companies were ordered to readjust contracts with advertisers so that advertisers would get back some money from the artificially-high rates paid under the monopoly. It is a remarkably strong judgment.

Before one thinks that the *Village Voice* case signals that the government may be more active in pursuing antitrust claims, it must be kept in mind how flagrantly the Sherman Act was flaunted here. If anything, the case suggests that corporations will estimate where the line is drawn between uncompetitive, illegal behavior and barely-legal, anti-competitive behavior. In *Village Voice*, they erroneously estimated where the line lay. But something made them think the line had moved in a very permissive direction.

Joint Operating Agreements

From about the end of the Civil War until the 1960s, most U.S. cities with large populations, typically 120,000 people or higher, had two or more daily newspapers with separate ownership competing directly with each other. Having two dailies had two major benefits. It forced competition in the advertising market, preventing imposition of extremely high rates. In the information and opinion markets, it assured more and varied coverage, as each competitor strove to increase or maintain circulation. Most Americans thought that having two newspapers to choose from was a norm.

From 1914 to the end of World War II, in midsize markets competition forced many competing dailies to close. And consolidation closed or merged many more. In the big cities, though, there remained many competing daily newspapers.

Closures in large markets began multiplying after 1945. Industry publications and academics began to address the issue, often with alarm (Meyer, 2004; Schmuhl and Picard, 2005). The reason for declining numbers of directly competing dailies was fairly simple, although the economics behind it were subtle. Once the number two paper fell far enough behind in circulation, it was unable to charge advertising rates that would let it stay competitive with the number one paper. Inevitably, the lost revenue would be reflected in less news coverage, which led to fewer readers. The cycle is called the circulation spiral (Lacy, 1987).

Ironically, by the time attention was focused on the problem, most cities had already become monopoly newspaper towns. The remaining two-newspaper cities were either very populous or were outliers where competition had continued despite economic pressures.

Ostensibly to assure or allow the continuation of competition in those communities that still had competing newspapers, Congress passed the Newspaper Preservation Act in 1970. The law allows two companies to merge the advertising, printing, and circulation units of their businesses so long as the two editorial operations remain separate. The contract to combine all operations except editorial is called a joint operating agreement, or JOA. Absent the exemption provided by the NPA, a merger of this type would likely violate the antitrust laws. It must be noted, though, that some pre-NPA partial mergers had been found legal under antitrust's "failing business" doctrine (Gillmor, Barron and Simon, 1998).

Under the NPA, two newspapers may petition to the Department of Justice for approval of a JOA when one of the papers meets the NPA's definition of a "failing newspaper." The term, borrowed from general antitrust law, represents a legal doctrine which allows a merger when one firm cannot return to profitability and there is no likelihood that a buyer can be found for that firm. In the daily newspaper business, though, deciding whether or not a newspaper might be able to return to profitability is far more complicated than it is with other types of goods and services. If one newspaper goes far enough down the circulation spiral to provide conclusive evidence that it is failing, it is also probably too far gone economically for its competitor to agree to a partial merger. The test under the NPA has become whether or not a newspaper has begun to fall into the circulation spiral or is on the edge of falling into it (Michigan Citizens, 1989; Committee for an Independent P-I, 1983).

An extra twist under the NPA is that the final approval decision is made by the U.S. Attorney General, making it harder to challenge because of administrative law's standard of deferring to agency judgments (Michigan Citizens, 1989). In most other areas of antitrust, action is taken to—or threatened to be—in court. In the *Michigan Citizens* case, members of the public opposed the JOA proposed between the *Detroit News* and the *Detroit Free Press*. The evidence was mixed. Both papers were losing money. Both papers had large corporate parents that had been subsidizing losses. But the *News* led in most indicia of competition, including circulation, and those indicia convinced the attorney general to not wait until the *Free Press* was worse off.

The approval review and subsequent court challenges in the Detroit case were lengthy. Later JOA request decisions were made more quickly (Department of Justice, 2000).

The cost to consumers of maintaining editorial competition is high. Once a JOA is approved, advertising rates jump up, less profitable circulation routes and areas are eliminated, and the overall news hole diminishes somewhat (Busterna and Picard, 1993).

The NPA has somewhat succeeded in serving its primary purpose of maintaining competing newspapers, but mainly by slowing down the rate at which cities have come to be served by only one newspaper. Since 1970, the number of JOAs has dropped from 25 to 12, even when the new JOAs are included. So the NPA has preserved limited competition in at least some cities. In the end, however, JOAs appear to delay the inevitable. A look at the demise of some JOAs helps explain why.

Dominant partners in two JOAs, each in very large markets, have recently sought to dissolve their agreements. This is typical. In general, most JOAs involve a morning paper and an evening paper, and the evening paper in the agreement is usually the one that continues in a downward spiral. But the JOA agreements normally guarantee each a share of the profits. Eventually the newspaper with higher circulation and higher revenues gets exasperated at having to subsidize the other, still-failing newspaper.

Seattle JOA

The *Seattle Times* is currently trying to dissolve its JOA with the *Seattle Post-Intelligencer*. The story of the Seattle JOA is representative. The two newspapers entered into a JOA in 1981 after getting approval from the Attorney General. Although the differences in circulation were slight, the *Times* led and the *Post-Intelligencer* had arguably entered the downward spiral. No evidence suggested that it could return to profitability. After protracted litigation brought a public interest group, the JOA was affirmed (Committee for an Independent P-I, 1983). At the time, the *Times* was owned by the Blethen family, the *Post-Intelligencer* by the Hearst chain. The *Times* was the evening newspaper. The 1983 Seattle court decision established that it was not necessary to prove that failure was inevitable. It also established that the finances of a parent company need not be considered in approving a JOA. To say otherwise would amount to requiring that chains subsidize their unprofitable properties.

The Seattle JOA agreement was amended in 1999. There were two key changes. First, the papers switched publication, with the *Times* becoming a morning newspaper. Second, Hearst was guaranteed a share of the profits until 2083, the new end date for the contract. But included in the contract was a provision that allows either newspaper to break up the partnership if it suffered losses during a three-year period. The *Times* invoked the provision in April 2003. At that time, the *Times* had 237,000 daily circulation, the *Post-Intelligencer* 151,000. Both papers had been the targets of strikes by two labor unions. In addition, a recession hit the Seattle area hard. Under the provision, the parties must either reach an agreement on how to separate within 18 months or, failing that, the JOA automatically ends.

Hearst filed suit in state court. A trial court judge held for Hearst, agreeing that a "force majeure" clause prevented the *Times* from using the three years of losses to invoke the breakup clause. A *force majeure* clause essentially prevents liability that results from events outside the control of a party to a contract. The Washington Court of Appeals unanimously reversed. It held that the two clauses were not linked in the contract. If the parties had wanted them linked, it must be in the contract. Hearst had relied on testimony and documents from the time period when the contract had been negotiated. Whatever each party may have wanted to say in the contract is not relevant, the court said. Unless the contract is ambiguous, and it is not, the language of the contract speaks for itself (Hearst Communications, 2004). In early 2005, the Washington Supreme Court agreed to hear Hearst's appeal (Richman, 2004).

The litigation appears to be part of a negotiation dance, however. The *Times* has stated that it really only seeks a "reasonable" renegotiation by Hearst, presumably one that guarantees a smaller share of future profits (Coughlin, 2004). The *Times* has publicly framed the dispute as an attempt by the Hearst Corporation to destroy competition in Seattle and to drown out one of the few independently-owned papers in the country (*Seattle Times*, 2005). The claim is more than a bit ironic. While the Blethen family still owns 50.5 percent of the paper, and family member Frank Blethen is publisher, the remaining 49.5 percent is owned by Knight-Ridder. Whether the JOA disappears or is renegotiated as the *Times* wants, the Blethen family and Knight-Ridder stand to achieve monopoly profits soon.

San Francisco JOA

The San Francisco JOA ended in 2000 for similar economic reasons but in a very different way. The *Examiner*, owned by Hearst, and the *Chronicle*, owned by Chronicle Publishing Company, had been in a JOA since 1964. Under the contract, profits were evenly split since their market position was roughly equal. In 1997, Chronicle announced it would not agree to an extension of the JOA. As a result, it would end no later than in 2005. After extended negotiation, Hearst agreed to buy the *Chronicle* and its half share of joint assets for about $660 million. Of that, the newspaper purchase was about $565 million. Chronicle sought bids from others. Knight-Ridder, Gannett, and Times-Mirror all bid, but not as high as Hearst. The two companies agreed that Hearst would buy the *Chronicle*.

The most unusual feature of this case is that the *Examiner*'s circulation had dwindled to about 115,000 from 301,000 in 1964, while the *Chronicle* had gone from 351,000 in 1964 to 482,000 in 1998. Usually the dominant paper takes over the market when a JOA ends. But Chronicle had unusual financial needs—its shareholders wanted to cash out, and that meant selling the *Chronicle* and other properties as well. Under the sale agreement, Hearst would make the *Chronicle* a morning paper.

Clinton Reilly, a local businessman and onetime candidate for mayor, filed a citizen's action to prevent the sale and dissolution of the JOA. In the case, a federal district judge upheld the parties' right to leave a contract. But the public's interest in the JOA was also noted (Clinton Reilly, 2000). To appease the Department of Justice, Hearst sold the *Examiner* to Florence Wang for a dollar (Department of Justice, 2000). Later Wang sold the paper to billionaire Phil Anschutz for $20 million (Harris, 2004). Anschutz continues to operate the *Examiner*, but not as a direct competitor of the *Chronicle*. Rather, it seems to have carved out a tabloid-style niche.

Honolulu JOA

In Honolulu, things turned out differently. Gannett owned the morning *Advertiser* and Liberty Newspapers owned the afternoon *Star-Bulletin.* Like afternoon newspapers in most markets, the *Star-Bulletin* had been losing circulation. At the time Gannett and Liberty sought to end the JOA, the afternoon paper had circulation just over 69,000 while the morning paper had circulation over 102,000. The *Star-Bulletin* was losing money. Since the Honolulu JOA contract guaranteed Liberty a certain level of profit, Gannett was bearing all the costs, even though those costs were allocated to the joint entity, the Hawaii Newspaper Agency. But the contract also had a provision for extra profits, profits Liberty could predict it would never see again. So the two companies negotiated a buyout of Liberty. The buyout would leave Honolulu with one paper.

The state brought an action to prevent dissolution of the JOA. Since the NPA by its own terms allows modification of any contract entered into before the NPA was passed in 1970, and the Honolulu JOA dated to the early 1960s, the defendants argued that there was no legal basis for the state's action (Gannett, 1999). The Department of Justice filed a brief siding with the state. Rather than endure protracted litigation, Liberty sought and found a buyer willing to operate the *Star-Bulletin*. The JOA was dissolved in 2001 and Honolulu continues to have two directly competitive daily newspapers. But one must wonder if it will last.

NPA Outcomes

The record on whether or not the Newspaper Preservation Act is desirable is mixed. While the policy goals—assuring two or more independent daily newspapers and keeping the public better-informed—are admirable, the NPA has served other goals as well, not all of them admirable. First, allowing JOAs lets large newspaper companies impose monopoly prices on advertisers with no concomitant attempt to assure that the public is served with a stronger overall news product (Fitzgerald, 2004). In the San Francisco JOA case, the trial judge essentially concluded that the two firms had milked profits while disinvesting in news. Second, the NPA operates on the quaint assumption that daily newspapering is more important than other ways of doing journalism (Gross, 2003). To turn that belief into a legal rule, though, is too great a step. It discriminates in favor of one medium and against others. In cases involving types of government benefits other than antitrust immunity, the U.S. Supreme Court has easily found First Amendment violations (Gillmor, Barron, Simon and Terry, 1996). It might see problems with the NPA if a case was framed well. Third, the NPA has done almost nothing to stem the tide of consolidation in the newspaper industry. Of the 24 remaining JOA partners, only six are independent companies. The NPA has proven a boon to conglomerates rather than a bane. And don't forget Seattle. Should the *Times* becomes the only daily, or the only meaningful daily, how long will the Blethen family likely wait to sell even more stock to Knight-Ridder?

Cross-Ownership Restrictions

The Federal Communications Commission attempted in 2003 to revise the rule which prohibited joint ownership of a newspaper and a broadcast station in the same market, abolishing it for many markets. The rule dates back to 1975, and had been upheld by the U.S. Supreme Court (Federal Communications Commission, 1978). The policy goal of the rule was to have a diversity of ownership, assuming that would produce more diversity of news and information. The decision also meant that existing cross-ownership situations would be grandfathered in. There were about 27 of those.

The rule was not a hard-and-fast one. Some exceptions had been made when separate ownership was no longer feasible (Gillmor, Barron and Simon, 1998). And buyers of media properties have even tried to argue that the cross-ownership ban should not apply to them even though they knew the rules in advance (Tribune Company, 1998).

Following its compulsory biennial review, the FCC decided to change the rule as a result of dramatic changes in the number and type of radio and over-the-air television stations in most communities. The cross-ownership rule change was just one of dozens of changes that were presented at the same time.

In making its decision, the FCC relied upon a commissioned study of ownership patterns of all media throughout the United States. It also looked at ownership patterns in broadcast and electronic media. Data from the study led to several conclusions. First, the fear that concentration of ownership may reduce the amount of information entering the marketplace of ideas, a key concern in establishing the 1975 rule, no longer applied in most markets. Thousands of new stations had been launched during the 28 years. Second, the FCC found that there were multiple alternative sources for news and information—mainly through the Internet and cable television—that did not exist before. Although the new media sources might not be substitutable, they provide a suitable safety valve. Third, the evidence showed that news product was more plentiful and of higher quality in some markets where a newspaper and broadcast property were jointly owned.

FCC Rule Change

The new rule was actually three rules. There would be no limits on cross-ownership in markets with more than eight TV stations, limited cross-ownership would be allowed in markets with four to eight TV stations, and no cross-ownership would be allowed in markets with three or fewer TV stations. The FCC concluded that concentration was still a threat to editorial vigor in the small markets.

The news that the cross-ownership rule was slated for extinction was met with enthusiasm by newspapers, at least by the large publicly-owned newspaper companies (Tompkins, 2003). The financial-advice-and-analysis firm Bear Stearns issued a study outlining how the new rules would effect various markets a full month before the rulemaking was formally announced (Bear Stearns, 2003). The Bear Stearns study delineated possible sales, purchases, trades, and acquisitions by the various large newspaper companies. Apparently the industry thought the rule change was a done deal.

Instead, the attempt to change the cross-ownership prohibition was met with an intense campaign by opponents of the new rule. Some opponents noted that the FCC had failed to account for the increased concentration of ownership in media overall in making its decision (Consumers Union, 2003). Other opponents focused on how the economics of the rule change could be detrimental to consumers. It would mean that a single company might eventually develop a monopoly status for local news and public affairs content. Given that the biggest buyer of TV stations in the past 10 years has been Gannett, the newspaper company, that looked like a reasonable projection (Gomery, 2002). Indeed, the Bear Stearns study pointed to markets where Gannett could develop desirable cross-ownership situations. Had Gannett been anticipating the rule change while buying up TV stations for years? The idea of a monopoly on local news and public affairs, even in medium and larger markets, is plausible. The local over-the-air television news business has developed its own version of the newspaper industry's circulation spiral. The TV news viewership spiral will lead to a single TV news product in many markets (Powers, 2001). Still other opponents focused on the methodology of the FCC's commissioned study. Eventually, public opinion became opposed to allowing cross-ownership (Pew, 2003), and with it in all likelihood the political will in Congress to support the FCC plan. How ironic: the Congress had decreed the biennial rules review, and in the 1996 Telecommunication Act it has voiced support for allowing cross-ownership.

Prometheus Decision

The rule changes were challenged in court. A Federal appeals court decided that the FCC had acted in an arbitrary and capricious fashion in setting up the new cross-ownership rules. The study was fatally flawed in numerous ways. In assessing the various media outlets in markets, the study had treated all equally. In other words, a public access cable channel was treated the same for editorial diversity purposes as was a broadcast network affiliate. The study also assumed that the availability of local web sites meant that there was more local news and public affairs information. The appeals court was especially irritated on this point, noting that supporters of the rule change were arguing that a city's web site was somehow substitutable. Finally, the three-tiered approach in the new rule was not supported by the data. The data did not show that cross-ownership would be good for large markets. And it did not show that cross-ownership would be bad for small markets. Under administrative law's rational basis standard, the FCC was found not to have a rational basis (Prometheus Radio, 2004). The court noted that the FCC does have the authority to dissolve the 1975 cross-ownership rule. The Supreme Court's 1978 decision upholding the rule does not require that the rule be continued. But if the FCC comes back with another change in the rule, it must have a better argument supported by better data.

Whether or not the Prometheus case means that the move to allow cross-ownership is halted is unclear. The appeals court practically invited the FCC to revoke the rule in toto since the three-tier approach was not supported. And the study did provide evidence that consumers might be better off, and certainly not worse off, if the cross-ownership ban was lifted. The point is moot for now in any event. The government dropped its planned appeal of the decision (Ahrens, 2005).

When the ban was put in effect in 1975, the empirical evidence did not provide strong support for the ban. In 2004, the empirical evidence did not provide strong evidence against the ban. In a sense, all sides of the issue are left to argue on theory and faith: that diversity in ownership is good because it should produce more and better content due to competition, or that concentration in ownership is good because it also should produce more and better content by being more efficient. The best guess is that the FCC and Congress, burned by the public relations disaster that the cross-ownership proposal became, will wait a good long while before revisiting this rule.

Conclusion

This chapter has reviewed government activity on regulation of newspaper ownership in the recent past. It found only a handful of cases where the government actually took any action, and in all but two of those, it chose not to do anything to prevent more concentration of ownership in the newspaper industry.

The two major antitrust laws are currently interpreted so narrowly by the courts and by the Department of Justice that only the most blatant anticompetitive activity will be allowed. Like the parties in the Village Voice case, one almost must write a memo that shows an agreement to fix prices to run afoul of the antitrust laws. But the newspaper industry is not an exception. Allowing consolidation has been the general industrial policy for over 20 years.

The Newspaper Preservation Act has provided financial life support for failing daily newspapers in two-newspaper towns. But even with life support, most of the patients have continued to weaken; eventually the JOAs get dissolved, with one company emerging as the winner. The NPA has assured more editorial voices, which is a positive, but whether or not the cost—in extortive advertising rates, jobs lost, and community uproar—have been worth it is uncertain. With only 12 JOAs remaining, reaching a definitive conclusion probably does not really matter any longer.

Curiously, the one area where there has been and continues to be actual regulation of newspaper ownership is in the continued prohibition of local cross-ownership of newspapers and over-the-air broadcast stations. But citizens groups are largely responsible for the failure of the government and big media companies when they tried to eliminate the ban, not government regulators.

References

Ahrens, F. (2005, January 28). FCC drops bid to relax media rules. *The Washington Post*, p. A1.

Antitrust investigations at NYT Co., Gannett (2005). Associated Press, reprinted January 24, 2005 in *Editor and Publisher*, p.

Bear Stearns and Co. (2003, June 2). 28-Year Newspaper-Broadcast Cross-Ownership Ban is Over. Author.

Beaudoin, C. and E. Thorson (2002). A Marketplace Theory of Media Use. *Mass Communication and Society, 5*, 241-262.

Berlyn, Inc. v. The Gazette Newspapers, Inc., Civil Action No. SW01-606 (D.Md., August 9, 2001).

Bridges, J. A., B. R. Litman, and L. W. Bridges (2002). Rosse's Model Revisited: Moving to Concentric Circles to Explain Newspaper Competition. *Journal of Media Economics, 15*, 3-19.

Busterna, J. and R. Picard (1993). *Joint Operating Agreements, The Newspaper Preservation Act and Its Application*. Norwood, N.J.: Ablex Publishing.

Chambers, T. (2003). Structural Changes in Small Media Markets. *Journal of Media Economics, 16*, 41-59.

Clayton Antitrust Act, 15 U.S.C. §§ 12, 13, 14-19, 20, 21, 22-27 (1914).

Clinton Reilly v. *The Hearst Corp*., No. C-00-0119-VRW (N.D. Cal., July 26, 2000).

Committee for an Independent P-I v. *Hearst Corp.,* 704 F.2d 467 (9th Cir. 1983).

Community Publishers, Inc. v. *Donrey Corp*., 892 F.Supp. 1146 (W.D. Ark. 1995).

Consumers Union (2004, January). Key arguments for rejecting the FCC's media ownership rules in the Feb. 11 Third Circuit Court of Appeals case. Available at consumersunion.org/pub/0503%20FCC%20petition%20to%20deny,%20AWE-Cing.pdf

Coughlin, K., (2004, November 30). Washington State Supreme Court will hear JOA appeal. Press release, *The Seattle Times*.

Department of Justice press release (2000, September 8). Antitrust Division recommends Attorney General approval of Denver newspapers' JOA.

Department of Justice press release (2000, March 30). Hearst Corp. to sell *San Francisco Examiner* to ExIn LLC.

Federal Communications Commission v. National Citizens Committee for Broadcasting, 436 U.S. 775 (1978).

Fitzgerald, M. (2005, January 31). Lee beats Gannett, employees for biggest Pulitzer Prize. *Editor and Publisher,* p.

Fitzgerald, M. (2004, November 3). At JOAs, the circ-rich get richer, and the circ-poor get poorer. *Editor and Publisher*, p.

Gannett Appeal Brief, State of Hawaii v. Gannett Pacific Corp., Case No. 99-17201 (9th Circuit, October 27, 1999).

Gannett press release (2005, March 31). Gannett completes acquisition of HomeTown Communications Network's assets.

Gillmor, D., J. Barron, and T. Simon (1998). *Mass communication law: cases and comment* (6th Ed.). Belmont, CA: Wadsworth, 544-547.

Gillmor, D., J. Barron, T. Simon, and H. Terry (1996). *St. Paul: West Publishing Co.*, 359-366.

Gomery, D. (2002) *The FCC's newspaper-broadcast cross-ownership rule: an analysis.* Washington, D.C.: Economic Policy Institute.

Gross, D. (2003, May 8). It's time to kill the antitrust exemption that allows rival newspapers to act like partners. slate.com Available at slate.msn.com/id/2082745.

Harris, R. (2004, February 20). Investor buys *San Francisco Examiner*. Associated Press, reprinted in Boston.com.

Hart-Scott-Rodino Antitrust Improvements Act of 1976, 15 *U.S.C. § 18a.*

Hearst Communications, Inc. v. Seattle Times Co., Docket No. 53148-4-1 (Washington Ct. App., March 22, 2004).

Ives, N. (2005, January 25) Justice Dept. Looks at Gannett Bid for Michigan-Based Publisher. *The New York Times*, p.

Jurkowitz, M. (2005, March 11). Globe completes Metro deal. *Boston Globe*, p.

Klein. J. (1997). *DOJ analysis of radio mergers*. U.S. Department of Justice.

Knutsen, K. (2005, March 6), *"One big newspaper,"* FightingBob.com. Available at www.fightingbob.com/article.cfm?articleID=344.

Lacy, S. (1987). The effects of intracity competition on daily newspaper content. *Journalism quarterly, 64*, 281-290.

Lacy, S. and T. Simon (1993). *The economics and regulation of United States newspapers*. Norwood, NJ, Ablex Publishing.

Lacy, S. and T. Simon (1997). Intercounty group ownership of daily newspapers and the decline of competition for readers. *Journalism and mass communication quarterly, 74*, 814-825.

Meyer, P. (2004). *The vanishing newspaper*. Columbia, Mo.: University of Missouri Press.

Michigan Citizens for an Independent Press v. Thornburgh, 868 F.2d 1285 (D.C. Cir. 1989), affirmed 490 U.S. 1045 (1989).

Myerson, B. (2005, May 2). Verizon boosts MCI bid to $8.5 billion. Associated Press.

Newspaper Preservation Act, 15 U.S.C. §§ 1802-1804 (1970).

Overture-Yahoo deal wins antitrust OK (2003, September 3). *Los Angeles Business*, p.

Pew Research Center (2003, July 13). Strong opposition to media cross-ownership emerges. Available at people-press.org/reports/display.php3?ReportID=188.

Powers, A. (2001). Toward monopolistic competition in U.S. local television news. *Journal of media economics, 14*, 77-86.

Prometheus Radio Project v. Federal Communications Commission, 373 F.3d 372 (3d Cir. 2004).

Pulitzer deal gets antitrust OK, company says. (2005, February 23) Associated Press. Reprinted in *Editor and Publisher*, p.

Richman, D. (2004, December 1) Justices to hear Seattle newspapers' dispute. *Seattle Post-Intelligencer*, p.

Schmuhl, R. and R. Picard (2005). The marketplace of ideas. In G. Overholser and K. Jamieson (Eds.). *The press* (pp.). New York: Oxford University Press.

Seattle Times (2005, May 5) To our readers. Available at seattletimes.com

Sherman Antitrust Act, § 1 et seq. (1890).

Sullivan, C. (2002, August 23). Antitrust case against Wash. Post Co. dismissed. *Editor and Publisher*, p.

Tompkins, A. (2003, May 29). *Cross-ownership, quality, and how the future may look.* Poynter online. Available at www.poynter.org/content/content_view.asp?id=35666

United States v. Times-Mirror Co., 274 *F.Supp.* (C.D. Cal. 1967), affirmed, 390 U.S. 712 (1968).

United States v. Village Voice Media, Civil Action No. 1:03CV1064 (N.D. Ohio, February 3, 2003).

With DOJ's antitrust OK, NYT Co. closes purchase of *Metro Boston*. (2005, March 11). Associated Press. Reprinted in *Editor and Publisher*, p.

In: Advances in Communications and Media Research, Vol. 3 ISBN 1-60021-188-7
Editor: Patula E. Grossmark, pp. 85-97

Chapter 6

STRATEGIC OPTIONS FOR THE NEWSPAPER PUBLISHING COMPANIES

H.L. (Hans) van Kranenburg*

University of Maastricht, Faculty of Economics and Business Administration, Department of Organization and Strategy, The Netherlands

ABSTRACT

The traditional publishing markets are undergoing a fundamental transformation. These developments pose threats to the traditional newspaper publishing companies, but also create new opportunities for these companies to engage in profitable new ventures and businesses. The companies have to find their way in the turbulence landscape. This study contributes to the existing analysis on the publishing strategy topic by putting the forces active in the market into perspective and by narrowing the problem down to such an extent that current and future changes in the publishing industry can be linked to possible strategic choices. A newspaper publishing company's competitive advantage for copying with the rapid changes in the landscape is not in its current distinctive competencies, but in those that it can grow tomorrow.

INTRODUCTION

The traditional publishing landscape has changed decidedly during the past decade. Technological and demographic developments, deregulation and the convergence of different

---

* H.L. (Hans) is an associate professor of Industrial Organization and Strategy at University of Maastricht, the Netherlands. His research interests concern industry dynamics, anti-trust issues, multimedia, information and communications industries and strategic behavior of firms. His work is published in e.g. Journal of Law and Economics, International Journal of Industrial Organization, Review of Industrial Organization, Managerial and Decision economics, International Studies of Management and Organization, Journal of Media Economics.

Contact address: University of Maastricht, Faculty of Economics and Business Administration. Department of Organization and Strategy. P.O. Box 616 6200 MD Maastricht The Netherlands, Tel: (31) 43-3883823; Fax: (31) 43-3884893. e-mail: h.vankranenburg@os.unimaas.nl information and communications markets have left an important mark on the configuration of the traditional publishing industry (Wirtz, 2001; Picard, 2003). Not only have these developments pose threats to the industry, but also create new opportunities for publishing companies to engage in profitable new ventures and businesses. While the performance and survival opportunities of traditional publishing companies, in particular the newspaper publishing companies, have been poor over the last decade it is expected that these companies will encounter serious problems to create strategies that will be able to benefit from the forces currently changing the traditional publishing landscape. Even if traditional newspaper publishing companies – on average- have been profitable in the last decades, their real competitive strength has been questioned. It is possible that the publishing company shows sign of above-average performance but do so without having a sustainable competitive advantage. It is likely that they have adopted the harvesting strategy. Some studies have pointed at the spiral of decline in the traditional publishing industry, in particular newspapers, due to destabilizing force of such factors as new entrants from outside the industry, innovations from existing information and communications companies and changes in customer behaviour. These developments are shown in many countries, for instance, in Argentina, Ireland, the Netherlands, North American and Nordic countries (Carroll, 1995; Sollinge, 1999; Kranenburg, Palm and Pfann, 2002). Newspaper publishing companies in these countries operate in more or less similar circumstances and under similar conditions.

The management team of these companies has had considerable difficulty in addressing the challenges raised by the rapid pace of current change in the traditional publishing industry. Declining profits because of increased competition and environment uncertainty have led to doubts about the newspaper publishing companies strategies' viability. Not only management, but also empirical evidence has failed to find conclusive answers on matters like scale and scope economies, alliances and consolidation benefits.

Newspaper publishing companies will have to find their way in the information and communications landscape. This information and communications landscape is a playing field larger than the traditional publishing environment. In order to survive and prosper in the today's changing industry structure and competitive situation, newspapers publishing companies clearly require a viable competitive strategy. They have to make adequate adaptations to these dynamic changes and respond quickly to create or to sustain their competitive advantage (Picard, 2003). This article contributes to the existing analysis on the publishing strategy topic by putting the forces active in the market into perspective and by narrowing the problem down to such an extent that current and future changes in the publishing industry can be linked to possible strategic opportunities.

Cost Leadership Strategy

What makes some newspaper publishing companies consistently more profitable than others within the same industry? Strategy issues for the newspaper publishing companies, which are closely related to the question, are why newspapers publishing companies differ, how they behave, how they choose their strategies and how they are managed. Of course, there exists a multitude of strengths and weaknesses that firms can have compared to their competitors. However, a firm can only posse two basic types of competitive advantage: cost leadership or differentiation (Porter, 1985). Other strengths and weaknesses are not less important but together determine either a cost leadership- or differentiation strategy.

Newspaper publishing companies pursuing a cost leadership strategy aim to become the low cost producer of information and communications products in the industry. These firms usually have a broad scope and serve many market segments. They may even operate in related industries, such as broadcast market (Ferguson, 1983) because this breadth of operations is often crucial in achieving cost-advantages. The way these cost advantages can be achieved depends on the structure of the industry. The cost structure of the publishing industry is characterized by a relative high ratio of fixed costs. The competitive cost advantages of publishing companies arises from the differences in capacity costs and high first-copy costs, like the collection and organization of editorial and advertising copy and preparation of the printing mechanism. These costs are ineluctable to produce even one product, while the marginal production costs of the subsequent products are very low (Picard and Brody, 1997). Ownership of two or more newspapers or ownership of broadcast stations and newspapers provides publishing companies opportunities for economies of size in news gathering, in securing advertising, in financing, and in management by lowering average and marginal costs. The costs advantages are greater for chain ownership, for high aggregate circulation, and for homogeneous products (Rosse, 1967; Norton and Norton, 1986).

Economies of scale are related to economies of scope, and these two terms are often used interchangeable. Economies of scope occur when a multi-product firm can produce given quantities of products at a lower total cost than the total cost of producing these quantities in separate firms. In other words, newspapers publishing companies have to ally parts of their value chain to benefit from savings across business activities. For instance, publishing companies can reduce the total administration costs when they have only one subscription network, instead of separate networks for their products. Another possibility is to outsource the subscription routines to independent companies, what already happens for magazines in Sweden (Hafstrand, 2002). Thus, an interesting option would be that the companies mainly concentrate on their core competence, collecting, processing and analyzing news, entertainment and information for their customers, while it can be argued that the other activities does not capture the core value creation logic of a newspaper publisher.

An important feature of a cost leader is that it will exploit all sources of cost advantage. In order to achieve a superior performance a cost leader must ensure that its products offering is of similar value to customers than that of its competitors. This is needed in order to translate the cost advantages obtained from using economies of scale and scope directly into higher profits. Newspaper publishing companies can achieve cost advantages with economies of scale and scope in the following areas:

- Information technology: This source may have great potential for publishing companies to achieve economies of scale and scope. The recent information technology developments facilitate an increase in efficient and effective utilization of information over ranges of products, services and customers. Information may allow for scope economies and facilitate competitive advantage to publishing companies that can offer a range of products and services to their customers. The possibility for reusability of information across customers has increased. Furthermore, the information technology developments may help facilitate differentiation of products and services (Chyi and Sylvie, 1998). New technologies

such as the internet and online subscription services make it possible to combine old and new media into one product with an additional element that was missing in the earlier markets: interactivity. The integration of activities allows a greater personalisation of relationships and interactivity between customers and publisher. The company can offer bundle of services and products into a complete service package for the customer (Wirtz, 2001). Newspaper publishers could offer services such as interactive forum, searchable news archives, online transactions, up-to-the minute information, and audio/video presentations. It is in the increased integration of activities and new technologies that lay the potential competitive advantage of publishing companies. Therefore, publishing interacts with many different fields within the information and communications industries and technologies. Because of this diversification publishing companies may have to redefine their ‘core business’. However to use these technologies in a competitive way, substantial investments in capital and expertise is required.
Distribution network: An important cost driver for the publishing industry, in particular the circulation industry, is distribution. For instance, distribution costs contribute about 20 percent to the total costs of a newspaper publishing company. See table 1. These companies can reduce their costs when they utilize more intensively their transportation network within a geographical territory, or when they generate the same volume of business in geographically smaller area as in a geographically larger area. These cost savings refer to economies of density. It is actually economies of scale along a given route that is reductions in average cost as volume on the route increase (Besanko, Dranove and Shankley, 1996). Of course, the recent technological developments may facilitate scale economies offering gains in running a greater sizable distribution network. The information technology enables companies to manage their network more efficient and faster. Hence, publishing companies can reduce the total cost of distribution when they have only one common distribution network, instead of separate networks for their products or they can outsource their distribution activities to independent companies.

Table 1. Development of the cost structure of Dutch newspaper publishing firms from 1968 to 1995 in percentage

	Year			
	1968	1975	1985	1995
Category				
Paper	21,0	20,5	18,2	12,6
Printing	18,0	24,2	19,4	17,5
Editing	16,0	19,7	20,3	23,6
Distribution	18,0	9,9	13,6	15,3
Advertising	5,0	6,1	6,2	7,0
Subscription	8,0	5,0	6,5	7,0
Buildings	14,0	14,6	15,8	17,0

Source: NDP annual reports

The distribution function of magazines in Sweden is already carried out by independent companies, although the largest distribution company, Tidsam, is owned and controlled by the larger magazine companies (Hafstrand, 2002).

- Production: Most of the existing studies on scale and scope economies focused on the potential production cost savings for traditional publishing companies. These studies showed the existence of economies of scale due to spreading product-specific fixed costs in circulation industries (Rosse 1967; Kranenburg et al., 2002). The production process of publishing products often involves high fixed costs. The product-specific fixed costs include the relative high expenses to set up a production process, for example, the time and expense required to set up a newspaper before printing. Newspaper publishing firms that produce the dominant design product in the market develop a competitive advantage. Hence, economies of scale in producing a (circulation) product are incorporated as a result of the possibilities of mass production of the product. Sufficiently strong scale economies can give large publishing companies enough cost advantages to force smaller and less efficient companies out of the market when the possibility of mass production increases or when the industry specific conditions become less favourable.
- Marketing and reputation: The next source of potential scale and scope economies is linked to marketing and reputation. It is generally understood that marketing may enlarge the readership of a newspaper (Picard, Winter, Mccombs and Lacy, 1988). Because marketing expenses involve substantial fixed costs, this implies the existence of economies of scale. Also, scope benefits may be present in the joint marketing of products to customers (Scherer and Ross, 1990). Another purpose of marketing is to create a brand image and loyalty of customers. Image and loyalty are related to the notion of 'reputation'. A company can build up a reputation with a consistent policy, marketing and highly qualified products and services. Reputation plays an important role in the publishing industry. Newspaper publishing companies offer products that crucially depend on their reputation and reliability. In general, customers relate closely the reputation and reliability of the publishing company to the quality of its products and services. Furthermore, the companies operate in industries that are characterized by the existence of a demand interdependence process. Circulation products are sold in two markets: the reader market and the advertising market. A change of a publisher's reputation in one market will affect the sells in both markets (Albarran, 1996). Increasing scale and scope allow publishers to capitalize more on their reputation. That is, a wider scope or scale may help a publishing company 'to put its asset 'reputation' at work'. For instance, an interesting aspect is that large companies and or companies with a very good reputation may have bargaining power to negotiate lower prices for their production inputs. Publishers can also use their reputation to diversify into number of unrelated and related businesses that are centered around their core business (Kranenburg, Cloodt and Hagedoorn, 2001).
- Innovations: Innovation in the publishing industry is an important source of scale and scope economies. Schumpeter (1934, 1942) argued that innovations give companies temporally monopolies because they have no competitors on the same playing field until a company duplicates or betters the innovations or products.

These temporally innovations are essential for the publishing companies because they can protect or increase the market share. Or even better they generate funds to strength their competitive position. For instance, the appearance of free newspapers has had a significant impact on the traditional newspaper market (Picard, 2001). The concept of free newspapers is based on only advertising income and free distribution in and around metropolitan public transportation systems. After the successful introduction of Metro in Stockholm incumbents and new entrants across the globe have begun launching also free newspapers. In general, the first movers in the geographically delimited markets had sooner or later to compete with other free newspapers. The RandD costs occurred by the first movers is likely to be higher than for the followers, because followers generally have to put less effort on the legitimacy of the innovation in the market. The existence of economies of scope and scale enables the first movers to spread the fixed costs of research and development over more customers. Innovations could be marketed to a larger consumer base and/or introduced in a wider set of activities. For innovations in the publishing industry scale and scope might be particularly important given the rapid speed of innovations and imitation by competitors. Only for a short period of time does a true competitive advantage exist. For instance, the introduction of the internet site by newspapers created a competitive advantage for those newspapers that introduced these innovations. In the Netherlands, the national daily newspaper NRC-Handelsblad was the first to enter the electronic newspaper market in 1995, shortly followed by its competitors, in particular the large newspapers. The larger newspapers had a higher opportunity to enter the electronic market than the smaller papers because of the needed investments. Increasing scale and scope allow publishing companies to recoup the fixed costs in this short period of time. Furthermore, large-scale companies can readily use internal capital reserves or raise funds on financial markets. Thus large publishing companies can start innovate activities almost immediately.

Differentiation Strategy

Publishing companies pursuing a differentiation strategy, seek to be unique in its industry along some dimensions that are widely valued by customers (Porter, 1985). They may seek one or maybe a few attributes that a large amount of customers find important and positions themselves uniquely to meet those needs. In other words, the differentiation strategy must add value in terms of greater customers satisfaction, lower cost, and unique features. When a newspaper publishing company offering a product and/or service or product and service group perceived by the customers as unique, the company is rewarded by a price premium. Of course, in the order to achieve superior performance a publishing company must ensure that the price premium earned exceeds the extra costs incurred for being unique. Differentiation is based upon firm-specific skills and creates a durable asset that is difficult to imitate or creates commitment of consumers. Therefore, it can create a sustainable competitive advantage for companies. For instance, it is difficult for many newspaper publishers around the world to imitate the image and reputation of specialized newspapers like the Financial Times and Wall

Street Journal. These newspapers are highly specialized in economic and financial information. Newspaper publishers also have to realize the importance of product differentiation in the on-line market. These products have to distinguish themselves from traditional media and also between other online products. The first group of differentiation refers to the development of internet-specific features such as a two-way communication, hypertext, and multimedia context that are not available in the traditional products. The differentiation between online products mainly has to focus on the information provided (Chyi and Sylvie, 1998, 2000; Hujanen, 2001)

The differentiating characteristics/attributes of publishing products can be divided into tangible and intangible features. The tangible features are observable differences that make the products or services better, cheaper or faster. For the publishing company this may include size, design features, material used, new technology that the product can now made of, consistency, speed of delivery, and accessories. The other group of features say something about the customer. The created "aura" of the product is important. The intangible aura of publishing products is one of image, lifestyle, quality and political, occupational, geographical or ethnical orientation. In short, the ability of a publishing company to implement a differentiation strategy crucially depends on the diversity of characteristics of consumers.

The impact that differentiation has on demand depends also on the competitive nature of the publishing product market environment. Two closely related concepts are important in the differentiation effort of publishing companies. These concepts are the stage of the product market evolution and the product life cycle (Hofer, 1975; Hill, 1988). The four stages of the two cycles are introduction, growth, maturity and decline phases (Vernon 1966; Gort and Klepper, 1982). These concepts develop simultaneously, however, when a product or product group may have its own life cycle within a broadly defined market. To formulate the strategy for a publishing company and also to determine its success, it is important to make a distinction between the structure of the market and the growth stage of the industry. Efforts to implement a differentiation strategy of publishing companies may be important in less fragmented mature markets, in particular established oligopolistic markets. In general, firms operating in these markets have achieved a minimum cost position and they can gain a sustainable competitive advantage only through some form of differentiation. As long as the products, services or product groups contain attributes to offer opportunities for differentiation than the differentiation strategy may be necessary to maintain current levels of demand. Hence, in these markets differentiation will not increase market share enough to enable the companies to realize substantial cost advantages, but to maintain its scale economies and safeguards its market share. However, the differentiation strategy can be successfully used to increase market share in emerging and rapid growth markets that are characterized by a low fragmentation. Under rapid growth conditions, entry of potential companies and differentiation to increase output will not necessarily reduce competitors' sales and decrease their profits. In these markets the incumbent firms will have less incentive to engage in aggressive response behaviour. Apart from the ability of the company to increase output, it can also realize scope and scale economies through differentiation. As long as the increase in costs due to differentiation is outweighed by reductions of costs associated with expanding output can differentiation be seen as a way of achieving a low cost-position. Differentiation efforts of companies operating in highly fragmented markets can also increase demand and help them to maintain a cost-advantage position. In general, the companies operating in high growth markets have the greatest opportunity to implement successfully a differentiation strategy to increase demand.

For newspaper publishing companies to maintain or create a sustainable competitive advantage, it is important to identify specific industry environments within which the cost advantage and differentiation strategy are likely to succeed. Efforts of differentiation in high growth markets and fragmented markets are more likely to realize economies of scale and scope. In slow or negative growth markets, as the printed newspaper market, a publishing company's expansion must take place at the expense of its competitors' market shares. In addition, through the increase in competition, companies are becoming more price and cost conscious, and as a consequence, they focus more on producing the dominant design product or product group to enjoy economies of scale and scope. Newspaper publishing companies can implement only successfully a differentiation strategy to increase output in mature markets that are fragmented or when it is possible to distinguish themselves again. The impact that the (publishing) product market environment has on the publishing companies' ability to differentiate is given in Table 2. This model developed by Hill (1988) is applied to the publishing industry. Publishing companies need to vertically integrate into the supply chain and or diversify into related and unrelated activities to implement successfully the differentiation and the cost advantage strategies in the near future. Product and international diversification is leading many publishing companies to pursue opportunities in the variety of new content providers that are developing and gaining significant public acceptance and also in the traditional businesses (Picard, 2003). It is obvious that publishing interacts with many different fields within the information and communications industries and their technologies, nowadays.

Table 2. The opportunity for newspaper publishing companies to implement a differentiation strategy successfully to increase demand given the product market environment

		Market Structure	
		High Fragmented	*Low Fragmented*
Stage of Industry Evolution	High Growth Stage	High opportunity e.g. web-sites; Exibitions; CD-roms and DVDs; direct marketing activities.	Medium opportunity e.g. internet provider; cable operator.
	Low Growth Stage	Medium opportunity e.g. leisure, professional and consumer magazines; Books; Scientific journals.	Low opportunity e.g. printed national and regional newspapers

Source: Hill (1988)

Strategy under Uncertainty

The problem of publishing companies is that they do not know what combination of activities will give them a competitive advantage in the future. In general, traditional newspaper publishing companies have become too dependent on their existing environment.

In their search for competitive advantages and higher profitability, it is not only necessarily important for these companies to enter new set of activities within new businesses but also exam the opportunity to enter traditional businesses that they traditionally choose to abstain from. Early entry, or better, choosing a wider set of activities would let the newspaper publishing companies discover which set of activities within the traditional and new businesses optimally fit together.

To create or maintain its competitive advantages, a publishing company obviously has two choices to deal with the new uncertain developments (Courtney, Kirkland and Viguerie, 1997).

One is to wait and see whether the market indeed offers the opportunities it thought it would. The publishing company should interpret the developments in the industry as a given fact. Hence, the publishing company will be a follower and have to rely on their flawless execution and imitation for competitive advantage. In particular, the adapter strategy makes sense for most small-sized publishing companies, because they do not have the financial resources and capabilities to explore the (potential) opportunities. This strategy's downside is, however, that the company may be too late to capitalize on the activity's profitability or that it has to make costly resource allocations it cannot afford anymore at that time. Therefore, the alternative option is to enter the new market early and find out whether the market does offer the expected profitability. This option permits the company to experience with a new business and to learn whether it has the capabilities and core competencies for this new business. When competition intensifies, the company has better knowledge about its own capabilities and market opportunities. The publishing company can expand in this business successfully, when it knows its capabilities are superior to those of its competitors and it should decide to exit the market when its capabilities are inferior. In the latter case, the publisher does not loose huge investments.

It is recommendable that the publishing company should use the scope expansion strategy to anticipate to future competitive environment for the new businesses. By following the scope expansion strategy a publishing company reserves the right to play in a variety of new activities. When the anticipated competition for the new activity is intensive due to for instance potential future competitors, early entry would be an optimal strategy. By making incremental investment and alliances today, the publishing company puts itself in a privileged position through acquisition of superior information by learning, development of a beneficial cost structure and creation of relationship with (potential) customers and suppliers. The strategy of publishing companies in the rapid changing environment may encompass several actions such as internal investment or to set up alliances, joint ventures, to acquire other companies and to divest activities.

VALUE OF MAJOR STRATEGIC OPTIONS

Newspaper publishing companies can internally developed activities that build on their competence in news, context and analysis. The decision to build independently on their traditional activities and looking for opportunities to create competitive advantages in a structural changing environment is a perilous route for the company. Publishers focusing only on internal development must frequently overcome technical constraints and legitimacy concerns in order to market products and services. Furthermore, the publishers often lack the financial funds, knowledge of the organizational and technical capabilities and skills that they require to create new businesses or to adapt existing businesses, in particular the small companies. However, in continue changing market conditions in the coming years, publishing companies need to begin to transform themselves into sustainable companies which rationalize current operations and adapt to changing technologies, markets, methods and competitive strategies (Picard, 2003). The absorption of at least part of the environment by means of mergers and acquisitions is one of the alternatives that publishing companies have if they attempt to reduce uncertainty, increase control over environmental changes, catch-up with new technological developments and trends or reduce their dependency on an existing environment (Hitt, Ireland and Hoskisson, 2002). A merger may help to create deep pocket so that investments in strategic options are more affordable. Furthermore, it may lead to diversification in capabilities. Of course, divestments may also be undertaken to finance some of the strategic options in which a company intends to engage. Companies must divest activities that are not part of the core-business anymore or activities which not generate the normal and necessary returns on investment. For instance, the company VNU divested its newspaper activities to finance the acquisition of AC Nielson Corporation in 1999. VNU has successfully transformed its company from a newspaper publishing company into an international information company.

A publishing company can choose to acquire the operating assets of another company in exchange for either cash, securities, or a combination of both. It acquires a stake in the other company. This means that the publishing company will have a certain degree of authority (or even full control) of what happens in the acquiring company. However, in some cases the publishing company may have difficulties in implementing the merger and acquisition strategy. For instance, the anti-trust authority can play a decisive role in the strategy of publishing companies. A merger or an acquisition by a publisher of another publishing company is generally not allowed when the company will obtain a too dominant market share. This is one of the reasons why alliances are an important alternative to mergers and acquisitions. This concept is underdeveloped in the context of publishing activities. This is to some extent surprising. The choice for an alliance can be a simple non-equity agreement – a cooperative alliance – to a partly equity transaction as a joint venture. The cooperative alliance is any contractual alliance between two or more industry incumbents or between companies that posses necessary complementary assets in which none of the companies have a degree of ownership, while a joint venture is a joint effort to achieve interests through the formation of a new entity by two or more business partners. Alliances provide access to a range of capabilities needed to develop focal and complementary goods, whereas the capabilities are often beyond the scope of any individual publishing company, in particular the small ones. The alliance strategy combines the benefits of an integrating (broader focused) publishing company and a stand-alone type of organization to maintain its current strategic position. In particular, it enables a publishing company to build a more-broader focused information and communications company next to running the relationship-rooted specific or regional publishing activities are prevented. In comparison to the internal development and acquisition strategy, the companies do not need deep pockets to implement the alliance strategy. In addition, the alliance strategy may enable in particular a large publishing company to act as a strategic player that can maneuver itself in a central position in between other companies. The company that is able to occupy a central position also has the potential to control the flows of information between these other companies. A central position in an inter-firm information and communications network provides the publishing company more insights in the dynamics of the markets that enables the company to use its visible position to further strengthen its position.

Of course, the alliance strategy does not come without cost (Garette and Dussauge, 2000). The important task for managers is to define a clearly defined portfolio of activities that would become part of the alliance, a portfolio of businesses to operate in and also the number of alliances. This is in particular important for small companies because they generally do not have the capability to manage a large number of partnerships and also many different forms of alliances. They must concentrate on just a few activities to improve on their strength in news, context, and analysis and have to be more careful in selecting their partner(s). Furthermore, existing networks may lock publishing companies into unproductive relationships, preclude partnering with other viable companies or may be unstable. Thus, a publishing company's network of relationships is a source of both opportunities and constraints. Because the landscape becomes sharply more competitive, the company's network enhances strategic importance and increases the survival opportunities of the publishing company.

CONCLUSION

Competitive pressure is forcing newspapers publishing companies to discover their competitive advantages, and choose an optimal configuration of services and activities. Due to rapid pace of changes in the traditional publishing industry the structure of the industry may be very different from the past. The current developments are just an interesting start.

To survive, publishing companies are forced to reconsider their strategy and product portfolio. A publisher's ability to change its organizational structure and existing activities and businesses along the evolution of technology in the industry can be seen as a major source of failure or success. This study helps managers understand how the competitive dynamics and company strategy are related. If managers have a better perspective of how their strategies are influenced by their context, they may examine the impact of such influence on the publishing company performance and be more conscious and aware of their company own strategic behavior. It is recommended that they focus more on the opportunities of partnerships in the near future.

Finally, a publishing company's competitive advantage for copying with the rapid changes in the landscape is not in its current distinctive competencies, but in those that it can grow tomorrow. A company's organizational ability to leverage and strengthen existing competencies is important, but it must be equally adept at diversifying its competence base. The bases for a publishing company's sustainable competitive advantage are its distinctive competencies. The strategic choices pursue by individual newspaper publishing companies will be influenced by a variety of local market conditions and internal organizational and operational factors. Although the strategies will differ among companies, they must continually emphasize and improve on their strength in news, context, and analysis. This is the core competence of these companies and also their competitive advantage.

Reference

Albarran, A. (1996). *Media Economics: Understanding Markets, Industries and Concepts*. Ames: Iowa State University Press.

Besanko, D., D. Dranove, and M. Shankley. (1996). *Economics of Strategy*. New York: John Wiley.

Carroll, G.R. (1995). 'Newspaper Publishers,' in Carroll G.R. and M.T. Hannan (ed.). *Organizations in Industry: Strategy, Structure and Selection. chapter 9*. New York: Oxford University Press.

Chyi, H. I. and G. Sylvie. (1998). 'Competiting with Whom? Where? And How? A Structural Analysis of the Electronic Newspaper Market,' *Journal of Media Economics*, 11 (2), pp. 1-18.

Chyi, H.I. and G. Sylvie. (2000). 'Online Newspapers in the US – Perceptions of Markets, Products, Revenue and Competition,' *International Journal of Media Management*, 2 (2), pp. 69-77.

Courtney, H., J. Kirkland and P. Viguerie. (1997). *'Strategy under Uncertainty,'* Harvard Business Review, 75 (6), pp. 67-79.

Ferguson, J.M. (1983). 'Daily Newspaper Advertising Rates, Local Media Cross-Ownership, Newspaper Chains, and Media Competition,' Journal of Law and Economics, 26, pp. 625-654.

Garette, B. and P. Dussauge. (2000). 'Alliances versus Acquisitions: Choosing the Right Option,' *European Management Journal*, 18 (1), pp. 63-69.

Gort, M. and S. Klepper. (1982). 'Time Paths in the Diffusion of Product Innovations,' *Economic Journal*, 92, pp. 630-653.

Hafstrand, H. (2002). 'Competitive Advantage in the Magazine Publishing Business: a Resource-based Perspective,' *Nordicom Review*, 2002 (1), pp. 291-300.

Hujanen, J. (2001). 'From Consuming Printed News to Making Online Journalism? Young Finns' Newspaper Reading at the Millennium,' *Nordicom*, 2001 (2), pp. 61-69.

Hill, C.W.L. (1988). 'Differentiation versus Low Cost of Differentiation and Low Cost: a Contingency Framework,' *Academy of Management Review*, 13 (3), pp. 401-412.

Hitt, M.A., R.D. Ireland and R.E. Hoskisson. (2002). *Strategic Management: Competitiveness and Globalization.* Cincinnatti, Ohio: South-Western College Publishing.

Hofer, C.W. (1975). 'Towards a Contingency Theory of Business Strategy,' *Academy of Management Journal,* 18, pp. 784-810.

Kranenburg, H.L. van, M. Cloodt and J. Hagedoorn. (2001). 'An Exploratory Study of Recent Trends in the Diversification of Dutch Publishing Companies in the Multimedia and Information Industries,' *International Studies of Management and Organization*, 31 (1), pp. 64-86.

Kranenburg, H.L. van, F.C. Palm and G.A. Pfann. (2002), 'Exit and Survival in a Concentrating Industry: the Case of Daily Newspapers in the Netherlands,' *Review of Industrial Organization,* 21, pp. 283-303.

NDP (Dutch Newspapers Publishers' Association) (1950-97). *Annual reports*. Amsterdam.

Norton S.W. and W. Norton Jr. (1986). 'Economies of Scale and the New Technology of Daily Newspapers: a Survivor Analysis,' *Quarterly Review of Economics and Business*, 26, pp. 66-83.

Picard, R.G. (2003). 'Cash Cows or Entrecôte: Publishing Companies and Disruptive Technologies,' *Trends in Communication* 11(2), pp. 127-136.

Picard, R.G. (2001), 'Strategic Responses to Free Distribution Daily Newspapers,' *International Journal of Media Management*, 2(3), pp.167-172.

Picard, R.G. and J.H. Brody. (1997). *The Newspaper Publishing Industry*. Boston: Allyn and Bacon.

Picard, R.G., J.P. Winter, M.M. McCombs and S. Lacy (ed.). (1988). *Press Concentration and Monopoly: New Perspectives on Newspaper Ownership and Operations*. Norwood, N.J.: Ablex Publishing.

Porter, M.E. (1985). *Competitive Advantage: Creating and Sustaining Superior Performance*. New York: Free Press.

Rosse, J.N. (1967). 'Daily Newspapers, Monopolistic Competition, and Economies of Scale,' *American Economic Review*, 57, pp. 555-577.

Scherer, F. and D. Ross. (1990). *Industrial Market Structure and Economic Performance* (3rd ed.). Boston: Houghton Mifflin.

Schumpeter, J.A. (1934). *The Theory of Economic Development*. Cambridge Mass: Harvard University Press.

Schumpeter, J.A. (1942). *Capitalism, Socialism and Democracy*. New York, N.Y. : Harper and Row.

Sollinge, J.D. (1999). 'Danish Newspapers: Structure and Developments,' *Nodicom Review*, 1999 (1), pp. 31-76.

Vernon, R. (1966). 'International Investment and International Trade in the Product Life Cycle,' *Quarterly Journal of Economics*, 80, pp. 190-207.

Wirtz, B.W. (2001). 'Reconfiguration of Value Chains in Converging Media and Communications Markets,' *Long Range Planning*, 34, pp. 489-506.

In: Advances in Communications and Media Research, Vol. 3 ISBN 1-60021-188-7
Editor: Patula E. Grossmark, pp. 99-114

Chapter 7

WEEKLY NEWSPAPERS IN THE MILLENNIUM

Hugh J. Martin
Grady College of Journalism and Mass Communication
University of Georgia, GA, USA

CHARACTERISTICS OF THE INDUSTRY

The economic health of weekly newspapers, unlike their daily counterparts, appears good. Weeklies are increasing in number and circulation as they enter the millennium (NAA, 2004). The number of dailies, and their circulation, is declining (NAA, 2004).[1]

Weeklies enjoy success despite a lack of professional respect. Scholars also give weeklies less attention than dailies. However, the lack of interest has not diminished weeklies' ability to operate in markets throughout the United States.

Many weeklies describe themselves as community newspapers, distinguishing their approach to news from the traditional detachment of many dailies. Weeklies publish information that strengthens "a shared sense of belonging" (NNA, 2004b) in the communities where they operate. The National Newspaper Association,, which has represented weeklies since 1885, reports its members serve communities based in geography, in politics, and in social or religious ties (NNA, 2004a). The Association also represents some dailies and Internet publications, but all members share a desire to strengthen communities (NNA, 2004a). Newspaper reading and strong community ties are closely intertwined (Hindman, 1998; McLeod, Scheufele, and Moy, 1999; Neuwirth, Salmon, and Neff, 1989)

Weeklies also have a significant presence in ethnic communities throughout the United States (Ma, 2003; Vargas and Pyssler, 1999; Zhou, 2002). Ethnic markets may be particularly attractive -- census figures show the number of Hispanics and Asians increased about four

---

[1] Daily circulation trends are complex (America, 2004). Total weekday circulation increased from 58,881,746 in 1960 to 62,766,232 in 1985. Circulation then declined, reaching 55,185, 351 in 2003.

times faster than the general population during the first three years of the millennium (Bernstein, 2004). Some daily newspaper companies are starting ethnic weeklies to take advantage of these trends (Truitt, 2001).

Two cost factors are also important for understanding the success of weeklies. The first is low entry barriers in weekly markets. A Georgia study (Martin, 2004) identified 86 new weeklies that entered the state's newspaper industry from 1990 to 2000. Starting a weekly costs less than starting a daily. For example, weeklies can rent expensive equipment such as a press, and jointly-owned weeklies often share a daily's production resources (Asher, 2002; Truitt, 2001). (Hugh –repeats later in text)

The second factor is free distribution of some or all copies of many weeklies. There is evidence (Stephen Lacy, Coulson, and Cho, 2002; Lacyand Dalmia, 1993) that some readers regard free local weeklies as complements to regional daily newspapers. Free circulation may also make it easier for new weeklies to acquire readers. For example, both Gannett and Lee Enterprises have started free weeklies, as well as websites, to attract younger readers with less interest in dailies (Eng, 2004; Johnson and Eng, 2004).

Weeklies are also adopting consolidation strategies similar to those used by daily newspapers to improve their competitive position. Like dailies, many weeklies are owned by groups (Asher, 2002; Guenther, 1987; Martin, 2004), which may allow them to take advantage of scale economies in both production and sales. Group-owned weeklies are also clustered in "geographically proximate" (Bridges, Litman, and Bridges, 2002) markets to increase their attractiveness to non-local advertisers.

This chapter expands on trends in the competitive market of weeklies and suggests the implications these trends have for the future of the industry. The first section describes models of weeklies within the newspaper industry. The second section discusses the economics of weeklies. This is followed by a section on the weekly's role within the communities where they operate. The last section focuses on current developments in the weekly industry.

Models of the Industry

Declines in direct daily newspaper competition inspired Rosse (1980; J. N. Rosse and Dertouzos, 1978) to develop a model of indirect competition between different layers of newspapers. The original model put weeklies, shoppers and specialty newspapers in the bottom of four layers of newspapers. The top layer of metropolitan newspapers provided regional coverage across neighboring cities and suburbs. The second layer was satellite city dailies focusing on their home city. The third layer was suburban dailies with local coverage. Weeklies could be in any geographic area, potentially facing competition from newspapers in any of the other layers. Additional layers of national dailies and group-owned suburban newspapers were later added to the model (Lacyand Simon, 1993).

Competition in Rosse's (1980; J. N. Rosse and Dertouzos, 1978) model can be based on circulation, on content, or both. Newspapers publish news and other information that attracts readers, then sell advertisers access to those readers (Lacyand Simon, 1993). Therefore, demand for readers and advertisers is interrelated.

The umbrella model described geographic segmentation in the industry (Bridges et al., 2002). Smaller papers reported local news to attract the readers that local advertisers wanted to reach. This was necessary because the small papers could not compete with cost advantages that metropolitan dailies enjoyed (Bridges et al., 2002). As a result, competition tended to flow down from higher to lower layers of newspapers.

However, changes in ownership and technology eroded the advantages larger papers enjoyed (Bridges et al., 2002). Those advantages included scale economies, which reduce long-run average production costs as more copies are printed, and sophisticated information about readers. Economies, and information about readers, are now within reach of smaller papers. Group ownership allows smaller newspapers to share production of news and advertising, further increasing scale economies (Bridges et al., 2002). There is evidence that placing group-owned newspapers in geographically adjacent markets is a particularly effective way to realize such economies (Martin, 2003a).

Bridges et al. (2002) modified the umbrella model to reflect these changes. Rings of newspapers were added to the original layers because papers can now compete within their own layer. Papers also can move within and between layers as they attempt to capture advertising, which has become more important than circulation (Bridges et al., 2002). The original specification (J. N. Rosse and Dertouzos, 1978) that weeklies, shoppers and specialized papers can compete in any market remains in the modified model.

Empirical Findings

Empirical tests of Rosse's model focused on dailies. However, some studies have examined the role of weeklies, finding complex patterns of competition. These studies include surveys and direct examinations of competition using circulation, content, and advertising.

Lacy (1984) studied competition at Southwestern newspapers in a survey of executives that included 56 weekly newspaper publishers. Executives at lower level newspapers did report competition flowed down from higher levels. However, weekly publishers reported that metro dailies primarily competed for advertising, not circulation (Stephen Lacy, 1984). The weekly publishers identified small dailies as their primary circulation competition. Inter-layer competition was called a moderate to severe threat to survival by 48.2% of the weekly publishers (Stephen Lacy, 1984).

A second study examined how the structure of a metropolitan paper's market influenced competition with weeklies and small dailies (Stephen Lacy, 1985). Markets where metropolitan dailies faced direct competition were compared with markets where the metro had a monopoly. There was no support for predictions that weeklies faced more intense competition from monopoly metros, which presumably had more financial resources and lower costs. Weekly publishers instead reported less advertising competition, and more circulation competition, from monopoly metros (Stephen Lacy, 1985). The study argued advertising competition may have been less intense because sales were limited by geography.

Correlations between content and circulation were examined in a study (Lacyand Sohn, 1990) that compared the substitutability of weeklies and metropolitan dailies in Denver and Detroit. Similar correlations at different newspapers would imply that content at one paper could substitute for content at another (Lacyand Sohn, 1990). However, most correlations were dissimilar. The exception was some advertising and sports coverage in Denver. This suggested that weeklies and metro dailies might be complements, not substitutes. If substitution occurred, it varied "from market to market and possibly from one type of reader to another" (Lacyand Sohn, 1990).

Two studies directly examined circulation competition between weeklies and dailies. Lacy and Dalmia (1993) studied competition in non-metropolitan counties in Michigan during four different years. Weeklies and dailies competed on weekdays, but not on Sunday. However, this study did not control for recursive relationships between layers of newspapers.

The second study (Lacy, Coulson and Cho, 2002) used two-stage regression to control for the recursive relationships between penetration in different newspaper layers. Newspaper competition was examined in a representative sample of 381 counties throughout the United States. The study excluded counties where metropolitan dailies where headquartered. There was weekday circulation competition between satellite dailies and weeklies (Lacy et al., 2002). However, when free weeklies were available, readers appeared to combine them with a subscription to a metropolitan daily to get a wider range of news (Lacy et al., 2002).

Two other studies examined advertising prices at weeklies. Blankenburg (1980) found weekly competition did not affect the open line ad rate. However, competition was measured by the distance between newspapers (Blankenburg, 1980).

A later study in Michigan (Lacyand Dravis, 1991) added advertising cost per thousand as a variable because the measure has been sensitive to competition in daily newspaper studies. Cost per thousand compares the price of reaching 1,000 subscribers at different weeklies. Lacy and Dravis (1991) measured competition as the number of weeklies in a county. The open line rate was not affected by competition . However, ad cost per thousand decreased as weekly competition increased (Lacyand Dravis, 1991). The study concluded weekly competition affected advertising prices in ways that were similar to daily markets.

Rosse's (Bridges et al., 2002; J. N. Rosse, 1980; J. N. Rosse and Dertouzos, 1978) umbrella model suggests a number of conclusions about these studies (Blankenburg, 1980; Stephen Lacy, 1984, 1985; Lacy et al., 2002; Lacyand Dalmia, 1993; Lacyand Dravis, 1991; Lacyand Sohn, 1990). Weeklies do face competition for circulation and for advertising from higher levels of newspapers, but that competition varies. Metropolitan dailies may compete for advertising that cuts across local markets where weeklies operate. However, smaller dailies compete for circulation because they are trying to capture more readers in markets where weeklies operate. Weeklies may also face direct advertising competition from other weeklies. Bridges et al. (2002) argued weeklies can respond to such threats by extending their own reach through the formation of groups. Longitudinal studies (Martin, 1999, 2004) in Michigan and Georgia have found increasing group ownership of weeklies, suggesting weeklies are forming groups to improve their competitive position.

However, the umbrella studies also suggest weeklies do not always compete with dailies (Lacy et al., 2002; Lacyand Dalmia, 1993; Lacyand Sohn, 1990). Some weeklies may instead complement metropolitan dailies, providing local news and advertisers to readers who then subscribe to the daily for additional content that is not available in the weekly.

ECONOMICS OF WEEKLIES: ENTRY AND DISTRIBUTION IN WEEKLY NEWSPAPER MARKETS

As mentioned earlier, weeklies may find it easier than daily newspapers to enter markets. The high capital costs of creating a daily newspaper are less likely to be a barrier for weeklies. For example, new weeklies might rent expensive equipment, such as contracting with a printer instead of buying a press. Weeklies can distribute themselves through the mail, avoiding the costly delivery networks required by daily newspapers. Guenther's (1987) study of Minnesota weeklies found many were printed on commercial presses. In the 1980s, a new weekly could be started for as little as $5,000 (Guenther, 1987). Weeklies are also sharing presses, content, or both in markets where the parent company also owns a daily newspaper (Asher, 2002; Lallande, 2002; Truitt, 2001).

The complementary co-existence of some weeklies and dailies may be explained by larger newspapers' tolerance for competitors who only serve readers and advertisers unwilling to pay the larger paper's prices. Co-existence may be tolerated by dailies because of scale economies in newspaper production. One *daily* newspaper can dominate a market because long-run average costs decline as production increases (Blankenburg, 1989; Bucklin, Caves, and Lo, 1989; Ferguson, 1983; James N. Rosse, 1967; Thompson, 1988). Competitors usually cannot capture enough market share to increase production to the point of matching the dominant paper's cost levels.

However, scale economies alone cannot deter the entry of new competitors over the long run (Caves, 1998; Scherer and Ross, 1990). Firms with scale economies may follow distribution policies that leave unserved portions of the market unwilling to pay their prices. This allows room for small firms, such as weeklies, operating on the market's fringe (Scherer and Ross, 1990). If the dominant newspaper pursues an aggressive price policy, firms on the fringe are likely to expand their market share as alienated readers and advertisers move away from the dominant paper.

An empirical test of these predictions compared weekly competition in counties with publicly-owned dailies *with? In?* counties with privately-owned dailies (Lacy, Coulson, and Martin, 2004). The test assumed that if publicly-owned dailies face more pressure to raise prices in pursuit of short-run profits, weeklies will have more opportunities to compete in those markets. The study of 292 randomly selected counties throughout the United States found publicly-owned dailies had more weeklies in their counties than privately-owned dailies (Lacy et al., 2004). An association between the number of weeklies and their household penetration suggested publicly-owned dailies risked losing more market share than privately-owned dailies (Lacy et al., 2004). However, the study did not directly measure price or spending policies at the newspapers involved. WE DID MEASURE HOUSEHOLDS.

Weeklies appear to fit well with a model developed by Caves (1998) predicting that new firms are likely to enter markets if capital costs are low. Low capital costs can then be viewed as a small, non-recoverable fee the new firm pays to discover if it operates more efficiently than incumbents (Caves, 1998). Firms that are more efficient will then reinvest using accurate information about the true cost of operating in the market. Inefficient firms can exit the market (Caves, 1998).

Two studies used this model to examine entry and exit in weekly markets. The first identified entering and exiting weeklies in Michigan during a four-year period in the late

1980s (Martin, 1999). The study identified 14 entering weeklies and 20 exiting firms. However, this was a pilot study and data were limited.

The second study (Martin, 2004) examined weekly markets in Georgia from 1990 to 2000. Eighty-six weeklies entered the industry, and 56 weeklies exited, accounting for substantial portions of total weekly circulation in Georgia. Entering weeklies were concentrated in suburban and metropolitan areas where most of the state's population growth occurred (Martin, 2004). The total number of weeklies in Georgia increased 16%.

The number of group-owned weeklies increased 220% during the 11-year study, and the number of individually-owned weeklies declined 23% (Martin, 2004). Group-owned weeklies distributed a higher percentage of free circulation. However, group-owned ad prices, measured as cost per thousand, were not consistently above the industry average (Martin, 2004). Martin (2004) concluded group-owned weeklies may hold down ad prices while distributing more free circulation to increase their competitive advantage in markets where entry barriers are low.

The research (Lacy et al., 2004; Martin, 1999, 2004) examining entry barriers in weekly markets suggests weeklies may not face the kinds of declines in circulation and numbers that dailies are experiencing so long as entry barriers remain low.

The Role of Weeklies in the Community

Attracting readers and advertisers is, of course, a problem facing all newspapers, not just weeklies. Readers are likely to include weeklies in a mix of different media only if weeklies add utility to that mix at a cost the readers consider attractive (Lacy, 2000). This is another way of saying weeklies will prosper if they are effective at differentiating themselves.

The industry's emphasis on weeklies as community papers that play a constructive role (NNA, 2004a) can be viewed as a differentiation strategy. This strategy is also recognized by individual companies, such as Community Newspapers, Inc., a Georgia-based firm that in 2003 owned 36 weeklies and three dailies in four Southern states (Community Newspapers, 2003). Community Newspapers promises in its mission statement to express "a bias in support of the communities we serve, providing aggressive, visionary leadership" (Community Newspapers, 2003).

The role that newspapers, including weeklies, play in their communities, has been examined in studies of relationships between newspaper reading and community involvement, of how news content reflects community issues, and of the function of ethnic newspapers.

A study of social relationships and media use concluded newspaper reading is among the forms of communication that play "a critical role in either motivating [community] participation or making it possible" (McLeod et al., 1999). This survey of 389 Wisconsin residents found a relationship between newspaper use and knowledge of local issues that was almost twice as strong as the relationship between television use and such knowledge (McLeod et al., 1999). Knowledge from newspaper reading in turn influences participation in political institutions, such as voting or contacting public officials (McLeod et al., 1999).

Another survey asked residents in 12 Midwestern areas about neighborhood involvement and the newspapers that respondents read (Jeffres, Atkin, and Neuendorf, 2002). Reading community newspapers was associated with attachment to a neighborhood, involvement in the neighborhood, and political involvement. Respondents who relied on neighborhood newspapers were less disillusioned with government, suggesting "grassroots" (Jeffres et al., 2002) print media had more impact on community involvement than other forms of communication.

Both studies (Jeffres et al., 2002; McLeod et al., 1999) suggest newspapers provide readers with information that fosters a sense they are part of, and involved in, a larger community. This information can prompt active participation in community affairs.

Relationships in the second study (Jeffres et al., 2002) changed as readers lived farther from a central city. Political involvement was associated with reading a daily newspaper in "more distant suburbs" (Jeffres et al., 2002). This may indicate metropolitan dailies cater to suburban audiences, making neighborhood papers more important in the central city (Jeffres et al., 2002).

An ethnographic study (Hindman, 1998) examined the role of one central city paper in a poor, ethnically diverse Minneapolis neighborhood. The paper fostered ties to the community with policies such as a promise to print all letters to the editor. Staff members were directly involved with important issues (Hindman, 1998). The paper's efforts to "represent the views of the entire neighborhood, to mediate conflict, and to promote neighborhood democracy" (Hindman, 1998) did encourage readers to become directly involved in local politics. The newspaper in this study was a monthly, but the findings seem relevant to weeklies.

However, an election-year study of letters to the editor in two New Jersey weeklies raised questions about their contribution to local political discussions (Sotillo and Starace-Nastasi, 1999). Rhetorical strategies in 123 letters were analyzed. Letters from political candidates tended to feature platitudes or details of the candidates' background. Ordinary citizens used letters to make highly polarized statements about local issues (Sotillo and Starace-Nastasi, 1999). There was little evidence of constructive dialogue, and many writers appeared to be alienated from local politics, the study concluded (Sotillo and Starace-Nastasi, 1999). The study did not examine other political content in the weeklies.

Ethnic communities may be a particularly attractive niche for weeklies given the rapid increase in ethnic populations. The number of Hispanics in the United States increased 13% between 2000 and 2003 (Bernstein, 2004). The number of Asians increased 12.5%. However, the overall population only increased 3.3% during this three-year period (Bernstein, 2004). Hispanics alone accounted for half of the overall population increase.[2]

Historically, ethnic newspapers in the United States arose in communities that were excluded from full participation in society (Diuguid and Rivers, 2000; West, 1995). The first of many African-American newspapers was started 1827 in reaction to coverage in white-owned newspapers (Diuguid and Rivers, 2000). Black newspapers were watchdogs for their communities, helped maintain black culture, and responded to attacks from white newspapers while offering a black perspective on issues such as slavery (Diuguid and Rivers, 2000).

The importance of African-American newspapers peaked from 1900 to 1954, when legal segregation began to end (Diuguid and Rivers, 2000). The civil rights movement also created opportunities for black journalists at mainstream white publications. "The black press still

[2] Estimates listed more than 290 million people in the United States in 2003 (U. S. Census, 2004). Eighty percent were white. About 37 million, or 12.8%, were African-American (U. S. Census, 2004). Almost 12 million, or

exists today, albeit with nowhere near the clout it maintained in the early 1900s. The positive side of this is that some of the country's leading newspapers include the names of black men" (Diuguid and Rivers, 2000).

White immigrants also created community newspapers during the late 1800s and early 1900s as a way of organizing responses to discrimination and repression (West, 1995). The number of ethnic newspapers in 77 United States cities increased from 171 in 1877 to 629 in 1914 (West, 1995). Newspapers only required small amounts of capital to start, but they helped ethnic communities organize democratic institutions that were critical to their assimilation into the larger society (West, 1995).

There still are ethnic communities where people may look to weeklies for this kind of civic support. For instance, ethnic media helped Chinese immigrants adjust to the United States by providing information in a familiar language (Zhou, 2002). Chinese media are usually owned by ethnic Chinese who employ immigrants to produce a mix of local, national and international coverage (Zhou, 2002). Weeklies have been an important part of the mix of ethnic media -- which includes dailies, broadcast stations, and Internet publications – that developed since Chinese immigration began increasing in 1965 (Zhou, 2002). Professional standards are not always enforced; for example reporters may receive favors from advertisers. However, the publications provide information about how to assimilate and succeed in the United States (Zhou, 2002).

Ma (2003) identified 40 Chinese newspapers, including three dailies, 12 weeklies, and 25 published less than once a week. Web sites allow the newspapers to reach readers in China and Taiwan. Thirty five of the newspapers had a web presence, including 10 of the weeklies (Ma, 2003). Ma (2003) argued the publications in the study have "played a major role in maintaining the identity, cohesiveness, and structure of Chinese Communities" (p. 179) in the United States. These publications are an important source of information about the "historical, social, political and economic development" (Ma, 2003, p. xx) of the communities they cover.

Another study (Vargas and Pyssler, 1999) asked if newspapers in Latino communities maintained "close connections with their readers, [to] teach them about the political process, and mobilize them for social change" (p. 192). The authors argued that community papers acquired by larger companies might focus on attracting advertising revenue and turn away from promoting social change.

Vargas' and Pyssler's (1999) content analysis examined health coverage in 4 Hispanic dailies, 12 weeklies and 1 monthly. Health received less front page play than other subjects, and stories overwhelmingly focused on symptoms and treatments. Stories rarely addressed social, environmental or political variables that influence health conditions. Many health stories lacked references to Latinos or community organizations, although weeklies were more likely than dailies to include such references (Vargas and Pyssler, 1999). The authors concluded "Latino newspapers are not realizing their potential as community media as effectively as they might" (Vargas and Pyssler, 1999, p. xx). However, the study did not measure newspaper ownership.

Studies of the role of newspapers in communities paint a mixed picture (Diuguid and Rivers, 2000; Hindman, 1998; Jeffres et al., 2002; Ma, 2003; McLeod et al., 1999; Sotillo and Starace-Nastasi, 1999; Vargas and Pyssler, 1999; West, 1995; Zhou, 2002). There is evidence

4%, were Asian. Another 39 million , or 13.7%, were Hispanics, who can be members of any race (U. S. Census, 2004).

weeklies can effectively differentiate themselves by consciously promoting community interests. Weeklies can also fill a niche by serving fast-growing ethnic communities. However, some of this differentiation may also require that weeklies operate with norms that differ from widely accepted standards of professionalism. West (West, 1995) also points out that newspapers which develop to organize a particular community are likely to vanish once their mission is complete.

Developments in the Weekly Industry

The standard newspaper reference first recognized the increasing importance of weeklies in 1999. The *Editor and Publisher International Yearbook* added a second volume (Gaye, 1999) devoted to information about weekly newspapers in the United States and Canada. The addition was explained by the editor: "In contrast to recent declines in the number of daily newspapers and daily newspaper circulation figures, the nondaily industry continued to show steady growth in each of these areas" (Gaye, 1999, p. xx).

The new listings defined weeklies as newspapers publishing 1 to 3 days a week that devoted at least 25% of their content to editorial material (Gaye, 1999) Weeklies were divided into categories based on how much of their circulation was distributed free. Paid weeklies charged for at least 95% of their circulation, but free weeklies distributed at least 95% of their circulation without charge. All other weeklies were defined as combination newspapers (Gaye, 1999).

A comparison of information in the *Yearbooks* for 1999 and 2004 (Gaye, 1999; Maddux, 2004) outlines developments in the weekly industry. Table 1 shows a net increase of 62 United States weeklies during this period, from 6,642 to 6,704. Total weekly circulation increased 1.5 million copies. Dailies are not shown in the table, but their numbers declined from 1,483 in 1999 to 1,456 in 2003 (NAA, 2004). Daily circulation declined from 55.9 million to 55.1 million during the same period (NAA, 2004).

Table 1 shows increases in the number of weeklies from 1999 to 2004 were concentrated among paid or free newspapers. Only about 17% of all weeklies were free, but they distributed almost half of all weekly circulation in the United States in 2004. Paid weeklies were the largest segment of the industry, and their number increased by 343 during the period shown. However, paid weeklies only distributed one-third of all weekly circulation in the United States in 2004.

Table 1 also shows a decline in the number and circulation of weeklies that distributed a combination of paid and free circulation. Combination weeklies were about 22% of the industry in 1999, but only 16% in 2004. Table 1 shows the number of combined weeklies declined by 378. The number of paid and free weeklies, however, increased by 440. The difference between these numbers accounts for the net gain of 62 weeklies from 1999 to 2004.

The changes in Table 1 suggest some combined weeklies may be switching to either paid for free circulation. Studies of Georgia and Michigan weeklies (Martin, 1999, 2004) suggest changes in overall numbers may also arise from entry and exit of weekly newspapers. Additional research is needed to determine more precisely what dynamics are reflected in the overall numbers.

Table 1. Recent changes in number and circulation of weeklies

Categories[a]	1999 Weeklies	1999 Circulation[b]	2004 Weeklies	2004 Circulation[b]
Paid	4,144	15.3	4,487	16.7
Percent	62.4%	31.5%	66.9%	33.3%
Free	1,040	22.5	1,137	24.5
Percent	15.7%	46.2%	16.9%	48.9%
Combined	1,458	10.8	1,080	8.9
Percent	21.9%	22.3%	16.1%	17.8%
Totals	**6,642**	**48.7**	**6,704**	**50.2**

Note: Based on data for relevant years in *Editor and Publisher International Yearbook*. Percentages may not sum to 100 due to rounding.

[a] Paid weeklies have at least 95 percent paid circulation. Free weeklies have at least 95 percent free circulation. Combined weeklies have both paid and free circulation.

[b] All circulation figures are in the millions.

Table 2. Specialty weeklies

Year	1999	2004
African-American	194	187
Percent	18.8%	19.6%
Ethnic	155	134
Percent	15.0%	14.0%
Hispanic	121	137
Percent	11.7%	14.4%
Jewish	106	103
Percent	10.3%	10.8%
Religious	160	125
Percent	15.4%	13.1%
Alternative	124	112
Percent	12.0%	10.8%
Gay and Lesbian	43	39
Percent	4.2%	4.1%
Military	131	117
Percent	12.7%	12.3%
Total	**1,034**	**954**

Note: Based on listings for relevant years in *Editor and Publisher International Yearbook*. Percentages may not sum to 100 due to rounding

The *Yearbooks* (Gaye, 1999; Maddux, 2004) also list specialty weeklies, including ethnic, religious, alternative, Gay and Lesbian and military newspapers. Table 2 shows that in 1999 there were 1,034 specialty weeklies, equal to 15.5% of the industry. Specialty listings declined in 2004 to 954, or 14.2% of all weeklies. There were declines in the number of specialty weeklies serving African-American, ethnic, and Jewish communities in the United States. The number of religious, alternative, gay and lesbian, and military newspapers also declined. The only increase was for Hispanic weeklies whose numbers rose 13% from 121 to 137.

Figures in Table 2 should be viewed with caution because many weeklies do not list themselves in the *Editor and Publisher Yearbooks*. However, Table 2 does raise questions about the opportunities for weeklies that fill specialized niches serving particular ethnic, cultural or religious communities.

Some of this may reflect changes in the structure of society. West (1995) argued that newspapers created in response to turmoil and repression in particular ethnic communities may vanish once those conditions change. Diuguid and Rivers (2000) argued that even though mainstream newspapers have not overcome a long history of racism, African-American newspapers began to decline as the civil rights movement began to succeed. West (1995) reported that newspapers were less important to organizing African-Americans than other institutions such as churches, historically black colleges, and the NAACP. Mastin (2000) recently found that churches and national media use were more important than local media for predicting civic involvement by African-Americans in one Midwestern community.

However, even if demand for some ethnic or other specialized weeklies declines, they seem unlikely to disappear. Specialized weeklies may provide important information for maintaining a community and its culture. Demand for such weeklies may also fluctuate. For example, many military weeklies are associated with a particular base, and reductions in their number may reflect base closings.

Increasing demand may explain why Table 2 shows the number of Hispanic weeklies increasing. As noted earlier, the number of Hispanics is increasing rapidly (Bernstein, 2004). Hispanics will have an estimated $686 billion in personal disposable income by 2009, according to one study (Humphreys, 2004) . This projection is 183% higher than the $242 billion that African-Americans, Asians, and Native Americans are expected to have available in 2009 (Humphreys, 2004).

Growth in the market for Hispanic weeklies may also reflect efforts by newspapers companies to reach Spanish-speaking immigrants (Truitt, 2001; Weeklies target Hispanics, 1999). Weeklies might also arise in some Hispanic communities if members feel isolated from the rest of society.

However, demand for Hispanic weeklies is likely to be complex. An experiment examined media use by Hispanics in one Southwestern community (Rios and Gaines, 1998). Television was preferred by non-English speakers who identified with their Latino heritage. English speakers with little interest in their heritage also lacked interest in newspapers. Only bicultural individuals who spoke *both* English and Spanish reported strong newspaper use (Rios and Gaines, 1998). Newspapers were used to help maintain cultural identity (Rios and Gaines, 1998).

The experiment (Rios and Gaines, 1998) was confined to 200 adults and cannot be generalized. However, it does suggest the Hispanic market is segmented, perhaps limiting opportunities for weeklies. Still, publishers continue to use weeklies as a way to tap into what is considered a growing Hispanic market (Johnson, 2004; Truitt, 2001; Weeklies target Hispanics, 1999).

Table 1 also shows that the national market for weeklies is slowly growing. There was an 8% increase in the number of weeklies and a 3% increase in circulation over 5 years. Much of the circulation growth also involved free weeklies, suggesting readers are not willing to pay for the content being provided.

Still, the continued growth and anecdotal evidence suggest weeklies are profitable. One newspaper broker predicted increasing interest in weekly acquisitions as the market for newspapers recovered from a downturn that began in 2001 and persisted for two years (Lallande, 2004a). Few dailies were being offered for sale, but weeklies were available. Increasing weekly circulation, and prices that were low compared with daily newspapers, made weeklies attractive to buyers (Lallande, 2004a).

Weeklies can capitalize on growth in suburban and rural areas (Nicholson, 1999), and newspaper companies are buying weeklies as a way of entering these markets (Lallande, 2004b). A former Gannett executive started his own newspaper company by purchasing seven weeklies and specialty newspapers in Alabama (Moses, 2000a). The low risk associated with community newspapers made them attractive to the executive and his financial backers (Moses, 2000a).

Daily newspaper companies are also buying weeklies in markets where the company already owns a daily (Asher, 2002; Lallande, 2002, 2004b). Dailies are increasingly interested in reaching segments of their markets that previously might not have been considered profitable enough to invest in.

Weeklies are also being consolidated under common ownership within and across geographically adjacent markets (Asher, 2002; Lallande, 2004b; Martin, 1999, 2004; Moses, 2000b). Studies of horizontal integration involving daily newspapers show clustering deters competition from other newspapers, creates new opportunities to reduce costs by extending economies of scale and scope, and increases market power (Fu, 2000; Lacyand Simon, 1997; Martin, 2003a, 2003b).

Weeklies are likely consolidating for similar reasons. Consolidated weeklies can be expected to share aspects of advertising sales, of production, and of distribution that will increase efficiency. In competitive markets, consolidation by weeklies may be necessary to ensure survival. In less competitive markets, consolidation may allow weeklies to raise advertising prices. However, research is needed before any of these effects can be established among weeklies.

CONCLUSION

Weeklies are a vibrant, important sector of the newspaper industry. There is no decline in the number of weeklies or their circulation. This may reflect the low entry barriers in weekly markets. Low startup and operating costs may leave weeklies well positioned to serve communities that dailies either overlook or have trouble reaching. These communities include specific metropolitan areas, ethnic communities, and readers with little interest in subscribing to a daily. The health of weeklies may also depend on their willingness to play an active role in community affairs.

For these reasons, weeklies appear well positioned to compete and even thrive in the new millennium. There are also signs the industry is adapting to changing economic conditions. Weeklies are becoming consolidated under group ownership. They are adopting clustering strategies so they can operate across adjacent markets.

Consolidation strategies can make weeklies more competitive by increasing their audience, allowing them to attract advertisers from outside their local markets. This strategy can also lower operating costs by creating economies of scale. The net effect would be to improve weeklies' ability to compete against dailies in other rings or layers of their markets.

However, the consolidation of daily newspapers in clusters has reduced competition in those markets, raised advertising prices, and resulted in the disappearance of many dailies (Fu, 2000; Lacyand Simon, 1997; Martin, 2003b). If weekly consolidation has similar effects, weeklies might also decline in some markets. Weeklies are also being acquired by their daily rivals (Case, 2000; Davis, 1999; Lallande, 2002, 2004b), which could also raise entry barriers in weekly markets.

Despite these challenges, there is every indication that weeklies will continue to be an important part of the newspaper industry. Larger trends in media use point to the increasing segmentation and fragmentation of audiences. Weeklies, already well established as niche players in an increasingly segmented industry, are well-positioned to capitalize on those trends.

The growing importance of the weekly segment of the newspaper industry also highlights the lack of knowledge about weeklies compared with daily newspapers. This is an area that will be deserving of attention from scholars and practitioners for many years to come.

REFERENCES

Asher, D. (2002, May). Welcome to Clusterland. *Presstime,* 29-32.

Bernstein, R. (2004, 2004, June 14). *Hispanic and Asian Americans increasing faster than overall population.* Retrieved July 25, 2004, from http://www.census.gov/Press-Release/www/releases/archives/race/001839.html

Blankenburg, W. B. (1980). Determinants of Pricing of Advertising in Weeklies. *Journalism Quarterly, 57*(4), 663-666.

Blankenburg, W. B. (1989). Newspaper scale and newspaper expenditures. *Newspaper Research Journal, 10*(2), 97-103.

Bridges, J. A., Litman, B. R., and Bridges, L. W. (2002). Rosse's model revisited: Moving to concentric circles to explain newspaper competition. *Journal of Media Economics, 15*(1), 3-19.

Bucklin, R. E., Caves, R. E., and Lo, A. W. (1989). Games of survival in the US newspaper industry. *Applied Economics, 21*(5 May), 631-649.

Case, T. (2000, April 17). Lee expands cluster. *Mediaweek, 10,* 28-30.

Caves, R. E. (1998). Industrial organization and new findings on the turnover and mobility of firms. *Journal of Economic Literature, 36*(December), 1947-1982.

Community Newspapers, I. (2003). *Community newspapers done right: A guide to publishing small-town newspapers* (2nd ed.). Athens, GA: Community Newspapers, Inc.

Davis, N. M. (1999, September). Ring Around The Metros. *Presstime, 21,* 51-58.

Diuguid, L., and Rivers, A. (2000). The media and the black response. *Annals of the American Academy of Political and Social Science, 569*, 120-134.

Eng, D. (2004, September). Old School, New Niches; Alternative papers go after younger readers. *Presstime,* 16.

Ferguson, J. M. (1983). Daily Newspaper Advertising Rates, Local Media Cross-Ownership, Newspaper Chains, and Media Competition. *Journal of Law and Economics, 26*(3), 635-654.

Fu, W.-J. (2000). *Three essays on the economics of networked and traditional information services.* Unpublished Dissertation, Northwestern University.

Gaye, M. G. (Ed.). (1999). *Editor and Publisher International Yearbook Part II* (79 ed.). New York: Editor and Publisher.

Guenther, L. (1987). 4 Categories of Change at the Weekly Newspaper. *Journalism Quarterly, 64*(4), 863-867.

Hindman, E. B. (1998). Community, democracy, and neighborhood news. *Journal of Communication, 48*(2), 27-39.

Humphreys, J. M. (2004). *The multicultural economy 2004: America's minority buying power* (No. Vol. 64 no. 3). Athens, Ga.: Selig Center for Economic Growth, Terry College of Business, The University of Georgia.

Jeffres, L. W., Atkin, D., and Neuendorf, K. A. (2002). A model linking community activity and communication with political attitudes and involvement in neighborhoods. *Political Communication, 19*(4), 387-421.

Johnson, T. D. (2004, February). Lost in Translation. *Presstime,* 17.

Johnson, T. D., and Eng, D. (2004, May). Free weeklies find a niche. *Presstime,* 20.

Lacy, S. (1984). Competition among metropolitan daily, small daily, and weekly newspapers. *Journalism Quarterly, 61*(3 autumn), 640-644, 742.

Lacy, S. (1985). Monopoly Metropolitan Dailies and Inter-City Competition. *Journalism Quarterly, 62*(3), 640-644.

Lacy, S. (2000). Commitment of financial resources as a measure of quality. In R. G. Picard (Ed.), *Measuring Media Content, Quality, and Diversity* (pp. 25-50). Turku, Finland: Media Group, Business Research and Development Centre, Turku School of Economics and Business Administration.

Lacy, S., Coulson, D. C., and Cho, H. (2002). Competition for Readers Among U.S. Metropolitan Daily, Nonmetropolitan Daily and Weekly Newspapers. *Journal of Media Economics, 15*(1), 21-40.

Lacy, S., Coulson, D. C., and Martin, H. J. (2004). Ownership and barriers to entry in non-metropolitan daily newspaper markets. *Journalism and Mass Communication Quarterly, 81*(2), 327-342.

Lacy, S., and Dalmia, S. (1993). Daily and weekly penetration in non-metropolitan areas of Michigan. *Newspaper Research Journal, 14*(3and4 Summer-Fall), 20-33.

Lacy, S., and Dravis, S. (1991). Pricing of advertising in weeklies: a replication. *Journalism Quarterly, 68*(3), 338-344.

Lacy, S., and Simon, T. F. (1993). *The economics and regulation of United States newspapers*. Norwood, N.J.: Ablex Pub.

Lacy, S., and Simon, T. F. (1997). Intercounty group ownership of daily newspapers and the decline of competition for readers. *Journalism and Mass Communication Quarterly, 74*(4 Winter), 814-825.

Lacy, S., and Sohn, A. B. (1990). Correlations of Newspaper Content with Circulation in the Suburbs - a Case-Study. *Journalism Quarterly, 67*(4), 785-793.

Lallande, A. (2002, November). Weeklies Extend Reach, Revenue; Dailies find success on a smaller scale. *Presstime,* 8.

Lallande, A. (2004a, February). 2004: Full Speed Ahead; Just 31 dailies changed hands in 2003. *Presstime,* 12.

Lallande, A. (2004b, November). Newspapers' Suburban Sprawl; Purchasing weeklies, shoppers key to ad growth. *Presstime,* 12.

Ma, Y. (2003). Chinese American Newspapers and Periodicals in the United States and Their Web Presence. *Serials Review, 29*(3), 179-198.

Maddux, D. (Ed.). (2004). *Editor and Publisher International Yearbook Part 2: Weeklies* (84 ed.). New York: Editor and Publisher.

Martin, H. J. (1999). *Factors Affecting Entry and Exit in Weekly Newspaper Markets: The Importance of Economies of Scale* (Unpublished). East Lansing, MI: Michigan State University.

Martin, H. J. (2003a). Clustered newspapers operate more efficiently. *Newspaper Research Journal, 24*(4), 6-21.

Martin, H. J. (2003b, July 30-August 2). *Some Effects from Horizontal Integration of Daily Newspapers on Markets, Prices and Competition.* Paper presented at the Media Management and Economics Division, Association for Education in Journalism and Mass Communication Annual Convention, Kansas City, MO.

Martin, H. J. (2004, Aug. 4- Aug. 7). *Characteristics of entry and exit in weekly newspaper markets: An exploratory study.* Paper presented at the Association for Education in Journalism and Mass Communication, Media Management and Economics Division, Toronto, ON.

Mastin, T. (2000). Media use and civic participation in the African-American population: Exploring participation among professionals and nonprofessionals. *Journalism and Mass Communication Quarterly, 77*(1), 115-127.

McLeod, J. M., Scheufele, D. A., and Moy, P. (1999). Community, communication, and participation: The role of mass media and interpersonal discussion in local political participation. *Political Communication, 16*(3), 315-336.

Moses, L. (2000a, April 17). A chain of one's own: Former Gannett exec makes it all happen. *Editor and Publisher, 133,* p. 14.

Moses, L. (2000b). Nashville Scene a bit blurry over move to chain ownership., *Editor and Publisher, 133*, p. 12: VNU eMedia, Inc.

National Newspaper Association (2004a). *About community papers*. Retrieved July 19, 2004, from http://www.nna.org/CommunityCenter/AboutCommNewspapers.htm

National Newspaper Association (2004b). *History of NNA*. Retrieved July 19, 2004, from http://www.nna.org/AboutNNA/NNAHistory.htm

Neuwirth, K., Salmon, C. T., and Neff, M. (1989). Community Orientation and Media Use. *Journalism Quarterly, 66*(1), 31-39.

Newspaper Association of America (2004). *Facts about newspapers 2004*. Retrieved November 16, 2004, from http://www.naa.org/info/facts04/

Nicholson. (1999, November 13). Newspaper money train to roll on, Veronis, Suhler predicts. *Editor and Publisher, 132,* 56-57.

Rios, D. I., and Gaines, S. O. (1998). Latino media use for cultural maintenance. *Journalism and Mass Communication Quarterly, 75*(4), 746-761.

Rosse, J. N. (1967). Daily newspapers, monopolistic competition, and economies of scale. *The American Economic Review, 57*(2 May), 522-533.

Rosse, J. N. (1980). The Decline of Direct Newspaper Competition. *Journal of Communication, 30*(2), 65-71.

Rosse, J. N., and Dertouzos, J. (1978, December 14-15). *Economic issues in mass communication industries.* Paper presented at the Proceedings of the Symposium on Media Concentration.

Scherer, F. M., and Ross, D. (1990). *Industrial Market Structure and Economic Performance* (3rd ed.). Boston: Houghton Mifflin.

Sotillo, S. M., and Starace-Nastasi, D. (1999). Political discourse of a working-class town. *Discourse and Society, 10*(2), 249-276.

Thompson, R. S. (1988). Product Differentiation in the Newspaper Industry - an Hedonic Price Approach. *Applied Economics, 20*(3), 367-376.

Truitt, R. C. (2001, November). Syndicates, en Espanol; Syndicated Spanish content multiplies. *Presstime,* 21.

U. S. Census Bureau. (2004, June 10, 2004). *Population estimates: National sex, race and Hispanic origin*. Retrieved July 25, 2004, from http://eire.census.gov/popest/data/national/tables/NC-EST2003-srh.php

Vargas, L. C., and Pyssler, B. J. D. (1999). U.S. Latino Newspapers as Health Communication Resources: A Content Analysis., *Howard Journal of Communications* (Vol. 10, pp. 189-205): Taylor and Francis Ltd.

Weeklies target Hispanics (1999). *Editor and Publisher, 132*, p. 17: VNU eMedia, Inc.

West, E. (1995). Organization Building in the Wake of Ethnic-Conflict - a Comparison of 3 Ethnic-Groups. *Social Forces, 73*(4), 1333-1363.

Zhou, M. (2002). Chinese Language Media in the United States: Immigration and Assimilation in American Life. *Qualitative Sociology, 25*(3), 419-441.

In: Advances in Communications and Media Research, Vol. 3 ISBN 1-60021-188-7
Editor: Patula E. Grossmark, pp. 115-126

Chapter 8

BUYING AND SELLING NEWSPAPERS

John Morton
Morton Research Incorporated
Silver Spring, Maryland, USA

The daily newspaper business is widely perceived as an "old economy" industry in decline. Its total national weekday circulation has dropped nearly continuously since 1987, down 12 percent since then, and its Sunday circulation down 6 percent since 1993. Yet daily newspapers continue to command high prices in the mergers and acquisition market. Individual newspapers, and not particularly large ones at that (the *Star-Tribune* in Minneapolis and *The Boston Globe*), have sold for considerably more than $1 billion in recent years. Entire airlines have sold for less. And large chains have brought even more -- $8 billion for Times Mirror Co. and $2.6 billion for Central Newspapers are examples. Why in the face of a declining presence and the emergence of myriad competitors, most recently the Internet, do newspapers remain so valuable? The answer is simple: profit. In 2003, not a banner year for newspaper advertising, the newspaper segments of the publicly owned newspaper companies collectively had an average operating profit of nearly 22 cents on every dollar of revenue. That's more than twice the average profitability of the Fortune 500 companies.

Not all newspaper properties are as profitable as those operated by the publicly owned newspaper companies, although profit is one key to what drives the continuing consolidation of newspaper ownership into even larger hands. There are other factors as well, and it is worthwhile exploring the forces that have contributed to ownership concentration in the newspaper business over the last 50 years or so. For while today's concerns about the impact of newspaper ownership concentration on the quality of public discourse seem urgent, they appeared no less urgent 50 years ago. As Emery and Smith wrote in their 1954 *The Press and America*, "Reduction of the number of American newspapers, and concentration of ownership of the remaining dailies, has caused a continuing debate over the social and political consequences of those trends" (p. 522).

The Sell Side

Family Ownership

Most daily newspapers in the United States, and probably all of those that still exist, began as essentially family enterprises operating local businesses. That was true of most other businesses, of course, but newspaper ownership tended to be different because newspaper owners, however imperfectly, always knew down deep that their businesses played a role in how well government and society work. Thus newspaper owners usually harbored emotional and even philosophical attachments to their properties not usually found among owners of other types of businesses. These attachments prevailed for the vast majority of newspaper ownerships even with the rise in the late 1800's of newspaper chains (which also were family controlled). By 1900, there were only eight chains controlling a total of 27 dailies. These chains collectively accounted for only 1.3 percent of all dailies and just 10 percent of national daily circulation (Mott, 1950). Today, 22 newspaper companies control 70 percent of the nation's weekday circulation and 73 percent of the Sunday circulation (Project for Excellence, 2004).

Clearly the attachment of families to their newspapers has waned. But despite the sharp difference between ownership concentration in 1900 and the present, ownership concentration came more slowly to the newspaper industry than to most other types of business, and family attachment likely was the major reason.[1] But over the last two or three decades other forces, chiefly the very nature of family ownership itself, have overwhelmed this pride in proprietorship. In addition, tax policies, the rise of publicly owned newspaper companies, profound changes in local retailing, and uncertainties about the future of newspapers have also been factors.

While the family head who founded a newspaper typically maintained firm control during his or her tenure, succeeding generations almost inevitably brought dispersal of shares in the company, not only to sons, daughters and grandchildren but also to cousins, in-laws and shirttail relatives. Many younger descendants of the founder might not have an active role in management and little sense of allegiance to the business. Thus, it was easy for some of these descendant share-holders to become disenchanted with the levels of dividends they received and, by extension, with what they perceived to be the inefficient management, profligate salaries and wasteful expense accounts of those family members who did run the company. A recent example of this dynamic involved Freedom Communications, owner of newspapers in Orange County, CA, Colorado Springs, CO, and numerous smaller cities. Dissident minority shareholders of Freedom almost succeeded in forcing the sale of the company; outside investors supplied sufficient financing for the loyal family members to buy out the dissidents. A variation of this theme involves members of a minority family in ownership stemming from an earlier consolidation of two newspaper companies into one. Minority family owners often are squeezed out of management over the years and usually have little or no say in management and dividend policy. Disputes within families and between controlling and minority families have often become so rancorous that the disputes wind up in court. Legend has it (no doubt apocryphally) that when the two families that owned the *Wichita* (KS) *Eagle-*

---

[1] Unless specifically noted, material in this chapter came from Audit Bureau of Circulations reports, *Editor & Publisher Yearbooks*, NAA *Fact Books*, and the Morton/Morton-Groves Newspaper Newsletters.

Beacon attended the annual company picnic, each side brought its own taster (now, as the *Wichita Eagle*, the paper is owned by Knight Ridder). Of course, much of the divisiveness within and between owning families could be avoided by a rational, equitable attitude toward such fundamental business practices as dividend payouts, compensation and perquisites of family members active in the business, and openness with financial data. These are the major sources of friction. Yet some newspaper families throw all they believe about business equity out the window when dealing with their own kin.

Often a consequence of these fractious relationships is the development of an over-the-counter market for a company's shares at local brokerage houses, where disgruntled shareholders can sell off their interests for cash, albeit usually at a low price relative to the company's true worth. This process spreads ownership even more, beyond the controlling family, and often to investors hoping to make a huge profit be pressuring the controlling family to sell out. Some or all of the elements cited above were in play in the sale of the Des Moines Register and Tribune Co., Cowles Media Co. (*Minneapolis Star and Tribune* and other properties), the Evening News Association (*Detroit News* and other properties), and the Courier-Journal and Louisville Times Co.

Estate Taxes

Another force driving the selling of newspapers is estate taxes. Many newspaper owners foresaw that their heirs very likely would have to sell the newspaper just to pay estate taxes, which could claim half the value of the family business. The heirs could borrow money to pay the tax bill, but interest on the loan could eat up much of the newspaper's after-tax income. A newspaper owner facing such a prospect for the heirs might logically ask: Why not choose before death which company will take over the family business? Selling before death means paying capital-gains taxes, at less than the half the estate-tax rate. The owner has a more intimate interest than anybody else in the family in the fate of the newspaper that, in effect, is the legacy of the owner's many years of nurturing. By selling before death, the owner can influence what kind of newspaper will live on simply by choosing the company to sell it to. Of course estate-tax pressures could be lessened by careful tax planning. Over time experts have devised strategies to protect family enterprises from much of the burden of estate taxes, including freezing values through recapitalization, cross-purchases and annuities, family gifts, charitable bequests, employee stock-ownership plans, and interest-free loans to children. But often by the time an owner realizes how compelling the estate-tax problem can be, it is too late to do much about it.

Immediate Financial Returns

While family dissension, squabbles with other minority shareholders, and estate-tax concerns are fairly concrete reasons for a decision to sell a newspaper company, another reason is somewhat amorphous and might suitably be placed under the rubric "the golf factor." For example, let's say the principal owner of a newspaper, usually a male but not always, has led a comfortable and prosperous life supported by a business largely free from the competitive pressures that create financial uncertainty for most business people. This owner typically has tended the newspaper's operations day-in and day-out for many years. Along comes a large media company with an offer to buy the family newspaper, and suddenly the owner realizes that all that separates him or her, and by extension other family shareholders, from the life of the idle rich is saying "yes." If the owner sells, he or she can stop working and spend winters in Florida playing golf and summers in Canada golfing and fishing (or substitute any other leisure-time interest). If the newspaper is large enough, he or she can choose Venice or Scotland or just about anywhere else on the planet as the playground. This can be a powerful incentive to sell, particularly if the owning family is apprehensive about the future of the newspaper business. If most or all of the family's assets are in newspapers, the family in recent years has to be aware that newspaper values are now as high as they have ever been. With the Internet looming, who knows what might be the case in five or 10 years. Therefore, the future holds an unknown risk; the present is worth a lot of money. This uncertainty likely was a factor in the controlling Chandler family's decision to sell Times Mirror Co. (the *Los Angeles Times*, *Baltimore Sun*, *Newsday* and other newspapers) to Tribune Co. for $8.3 billion in 2000 and, later that year, for the Pulliam family's decision to sell its Central Newspaper (*Arizona Republic* in Phoenix, The *Indianapolis Star* and other newspapers) to Gannett for $2.6 billion. Indeed, the Pulliam family's trust supposedly was unbreakable and assured continued family control until late in the 21st Century, but a lawyerly phrase in the trust documents concerning unknown future exigencies apparently was enough to break the trust and permit the sale.

Clustering Behavior

Finally, a new factor encouraging sales of newspapers, especially smaller ones, emerged in the mid 1990's, occasioned partly by changes in retail markets and partly by the quest for economic efficiency. The new factor is known as clustering. For small dailies in rural markets, the increasing consolidation of small family farms into large, heavily mechanized agribusinesses has sharply reduced farm populations. Fewer people come to town to buy groceries, clothes, insurance, feed and seed, farm implements, and the other goods and services that drive local rural economies—including, of course, the local newspaper's advertising revenue.

Moreover, these markets, as well as those not crucially dependent on agriculture, also have gone through changes in retailing that have adversely affected newspaper revenue. The best known of these changes is called the Wal-Mart effect. This effect takes place when a big discount retail store moves into small market, advertises in the newspaper heavily at first, then withdraws advertising as small locally owned retailers either give up or become so financially hard-pressed that they slash their advertising budgets. With little significant competition left, the large discount store no longer needs to advertise to pull in customers.

Even in those markets large enough to support more than one discount store and in which some advertising competition remains, individually owned newspapers are hurt by the trend toward regional or national ownership of local retail stores; this trend has diminished the presence of locally owned stores and has meant that the same chain stores turn up in market after market. Before this retail consolidation, locally owned advertisers and local outlets of chains like Sears made advertising spending decisions locally. Now these decisions are made at regional or national offices. Increasingly, these regional and national offices prefer to advertise with newspaper companies that own newspapers in several, rather than one, of the markets in which the chains have outlets. A locally owned daily is in danger of dropping off the map when advertising decisions are made this way. In turn, this trend of advertisers' wishing to cover several markets with one buy has contributed to geographic clustering of newspaper ownerships, as newspaper companies seek to expand their appeal to chain advertisers. And, of course, this clustering of ownerships adds pressure to remaining singly owned newspapers to sell out to a larger company's cluster.

THE BUY SIDE

Taxation Policies

The foregoing explains the myriad forces that over the years have contributed pressure on newspaper owners to sell out to a larger company (the sell side). There likewise have been various pressures on the buy side of the equation. Newspaper companies, especially large ones, throw off a lot of profit. This gives these companies essentially two choices as to what to do with all the money: Pay dividends to shareholders or use it to buy something else. Because of double taxation, boosting dividends often is not attractive, especially if the benefiting shareholders are mostly family members. In other words, a newspaper company pays tax on its profit and some of what is left over can be paid as dividends; dividend recipients in turn have to pay tax on the dividends. Leaving dividends alone and just letting profits pile up will not work for long, because these profits could trigger the accumulated earnings tax. In effect, this tax is the government's way of capturing what it would have received in taxes if dividends had been paid. It should be noted that, compared with earlier times, recent tax reductions enacted by the Congress have considerably lessened the impact of both double taxation and the accumulated earnings tax (Jobs and Growth Tax Relief Reconciliation Act of 2003 Public Law 108-27). The rate on dividends and the accumulated earnings tax are now down to 15 percent. Still, over the years double taxation and the accumulated earnings tax meant that newspaper companies historically have maintained relatively low dividend payouts in favor of investing profits in acquiring additional newspaper properties (some companies as well have invested in broadcast, cable, and other operations). In effect, tax policies have encouraged newspaper companies to become larger.

Chain Growth

Publicly owned newspaper companies, in particular, have sought to become larger. When newspaper companies started going public in the 1960's and 1970's, their managements quickly realized that institutional investors, which dominate the public markets, reward growth. When a steel company or a green-bean canner has extra money, it can choose to build a new plant to increase growth if it cannot find one to buy from somebody else. Rich newspaper companies usually do not have this choice. Every market large enough to support a daily newspaper already has one, and newspaper companies learned long ago that it is hopeless to try to start up a competing daily in a market already being well-served by an existing one. Creating a competing daily takes decades of investment and losses without any assurance of success. This is not a formula for growth but rather for a decline in profitability. There have been a few exceptions to the reluctance to start up new dailies. *USA Today*, after many years of losses, found a profitable niche in the national newspaper market, but *USA Today* and its niche are atypical of the daily newspaper industry as a whole. From time to time daily newspapers have been created in local markets – recently *The New York Sun* and earlier *The Washington* (D.C.) *Times*, The *Sun* is too young to pronounce it a success, but its small circulation (40,000) and scant advertising are not promising. The *Washington Times* is subsidized by the Unification Church and has acknowledged losing several millions of dollars every year.

Exceptions of more importance involve existing, dominating dailies starting up new newspapers in their own markets to seek out ethnic or other special audiences. In effect these are investments in trying to expand the dominant dailies' franchises. There are now free or cheap dailies in four cities aimed at commuters and specifically at the 18-to-34 age group, all started by what might be called "mother" newspapers. Examples are the *Chicago Tribune*'s *RedEye*, the *Chicago Sun-Times*'s *RedStreak*, *The Washington Post*'s *Express*, and *The Dallas Morning News*'s *Quick*. Gannett and other publishers in smaller markets have created weeklies with the same orientation. These start-ups reflect the newspaper industry's recognition that there is a large and growing portion of the population -- young people -- that is relatively well off but that has not taken up newspaper-reading in nearly the numbers of former days. The hope is that these youth-oriented endeavors, with their flashy, entertainment-oriented and greatly abbreviated journalism, will entice younger readers into developing a daily-newspaper reading habit that eventually will be transferred to the more serious mother papers. It is the failure of traditional daily newspapers to attract young readers that is the primary cause for the newspaper industry's circulation declines since the late 1980's. In effect, newspaper readers are dying off faster than they are being replaced.

The effort to establish dailies aimed at ethnic audiences has a somewhat different focus. So far, these endeavors have involved creating Spanish-language dailies. Examples are the Tribune Co.'s *Hoy* newspapers in New York, Chicago and Los Angeles and The Dallas Morning News's *Al Dia* and the Fort Worth Star-Telegram's *La Estrella*. All of these efforts are tied to mother papers except for Tribune's *Hoy* in New York City, which is not closely linked to the company's *Newsday* on Long Island. The strategy in creating Spanish-language dailies is founded on the fact that Hispanic immigration is rising much faster than assimilation of Spanish speakers into the English-speaking mainstream. Waiting for assimilation to turn recent immigrants into readers of English-language newspapers could cause the newspapers to risk losing a substantial number of potential readers.

Economies of Scale

Aside from the exceptions mentioned above most newspaper companies seek growth through acquiring existing newspapers. Nor is simple growth in size the only motivation for this strategy. Taking over a locally owned family newspaper offers a large newspaper company many opportunities to improve profits. A company like Gannett buys everything from newsprint to paper clips at prices cheaper than those paid by a single newspaper, since Gannett buys in big volumes and a single newspaper does not. Moreover owners of individual local newspapers do not have as a primary goal showing strong profits, which brings a higher tax bill. Instead, payrolls are often bloated by the presence of family employees, who prefer to take money out of the company in the form of wages that are taxed once rather as dividends that are taxed twice (as noted, the recent reduction of the dividend tax to 15 percent might lessen this preference). And typically there are other perquisites for family employees – such as company cars and travel costs. When a large newspaper company buys a newspaper there usually is a paring, if not outright elimination, of the family payroll and perquisites. Some of the administrative costs previously borne by the acquired newspaper also are taken over at the corporate level of the acquiring company, further reducing costs. It is not unusual for reductions in costs at the acquired newspaper to double the newspaper's profitability over two or three years, which of course contributes to the earnings' growth of the acquiring company.

Shareholder Liquidity

The desire to achieve growth through acquisitions is felt especially at the 15 publicly owned newspaper companies, which collectively control 52 percent of both weekday and Sunday circulation in the United States (Editor and Publisher 2003 Year Book, based on Audit Bureau of Circulations and other data). Large, family-controlled newspaper companies began moving to public ownership in noticeable numbers in the 1960's. The reasons given for these companies' moving to public ownership were various – to pay off debt, the ability to use publicly traded shares to make acquisitions, using proceeds to invest in new presses and technology. But probably the most important was to give existing shareholders liquidity – the opportunity to sell their interests in the company in an efficient market. This liquidity is also a mechanism that can serve to get rid of dissenting family shareholders and other minority shareholders unhappy with management.

While the conventional wisdom is that once a company is publicly owned it is truly owned and controlled by shareholder votes, this was really not the case with a majority of the newspaper companies going public. Instead, most companies devised a reorganization that involved two classes of stock – one super-voting that would be retained by the family and would not be traded and another either non-voting or weak-voting that would be offered to the public. Moreover, most of the plans provided that if holders of super-voting shares decide to sell their interests, their shares must first be converted to the publicly traded non-voting or weak-voting shares. Thus did the dominant family owners assure themselves of continued control, immune to tender offers from outsiders. The company would be sold only when the controlling family shareholders decided to sell. Some companies, notably Gannett, Knight Ridder and Tribune, did not opt for this stock structure. Another company than did not was Booth Newspapers, which ultimately was acquired by Advance Publications (Newhouse) through an unfriendly tender offer. Eight newspaper companies had public offerings in the 1960's, seven more did in the 1970's, and seven more in the 1980's. Not all of these companies remain as publicly traded entities. Knight Newspapers and Ridder Publications merged into Knight Ridder in 1974; Gannett acquired publicly traded Speidel Newspapers, Multimedia, Inc, and Central Newspapers; Tribune Co. acquired Times Mirror Co.; E.W. Scripps Co. acquired Harte-Hanks Inc.; Media General acquired Park Communications, and a couple of other, smaller public companies were sold off piecemeal. These mergers and acquisitions of course contributed significantly to ownership concentration.

Short-Term Profits

Institutional investors were attached to the stocks of these publicly owned newspapers, recognizing that most of the newspapers owned by the companies operated as unregulated monopolies capable of throwing off profits at two to three times the rate of most other businesses. In the early years of newspaper public ownership, many institutional investment firms, which account for a majority of the trading on the major stock exchanges, tended to invest for the long term. In effect, they were investing in the future of a company. This investment approach began to wane in the 1980's and practically disappeared in the 1990's. It was replaced by a strategy focused much more on the short term, with much more buying and selling of stocks. This meant that for purposes of bonuses fund managers increasingly were judged on the annual or even quarterly profit performance and stock-price appreciation of the companies they had invested in. This put a premium on short-term profit growth, and what the financial community cares about is quickly translated into pressure on the publicly owned companies to care about the same thing. The financial community, being what it is, now primarily concerns itself with financial performance. This is a group that often has a natural inclination to think of an editorial staff as a non-revenue-producing controllable expense and, sadly, under pressure some newspaper companies began to think the same way. At least since World War II until the early 1990's, it was almost unheard of for a newspaper company to lay off employees when a soft economy threatened profits. The publicly owned newspaper companies found themselves with a new constituency in addition to readers, one not easily convinced of the need actually to increase editorial spending in the face of rising competition and costs and at the expense of profits. The newspaper industry's much-lauded high profitability had established a level of expectation about financial performance, and this in turn increased pressure on publicly owned companies not only to engage in cost-cutting but also to step up acquisitions as a method of increasing earnings. Indeed, even privately owned newspaper companies with ownership dispersed after three or four generations found themselves subject to the same pressures.

Property Movement

The convergence of the pressures on the newspaper owners to sell out to larger companies with the pressures on large companies to acquire additional properties for earnings' growth reached a peak in 2000. In dollar terms, transactions that year, in which 125 dailies changed hands, had an estimated total value of $14 billion (Dirks, van Essen and Murray, 2001). Because of the onset of the recession in 2001, the number of dailies changing hands dropped sharply, to 23 in that year, and rose only slightly through 2003. Still, there remain many properties to be acquired. Of the 1,456 dailies that remain in the United States, only about 260 remain locally owned, and most of these have small circulations. The rest are owned by 120 companies that own two or more dailies. Circulations of many of these companies are quite small – 53 of them own only two to four dailies, typically small in circulation. In effect, these are family companies that have expanded from an original base of one newspaper, and they will be acquisition targets not much different from the singly owned dailies. Of the 67 remaining companies that own five or more dailies, many of these, too, are collections of relatively small newspapers, but others include the largest newspaper companies in the country, some with vast holdings. Among larger ones are Advance Publications (Newhouse), Cox Newspapers, Dow Jones and Co., Gannett Co., Tribune Co. plus two or three others. These companies are not likely targets for acquisition; rather they are likely to be acquirers. Removing these companies as potential acquisition targets leaves approximately 110 newspaper chains and the 260 or so singly owned dailies.

Selling Chains

During the 1970's and 1980's, the sale of a newspaper company usually meant that the owning family of a single newspaper sold out to a larger company. During the 1990's, though, it was the sale of newspaper chains themselves that became notable. After decades of buying single family-owned dailies, there were not many independent dailies left that were of a large enough size to be attractive to the largest chains. However, single newspapers were still being sold, usually to small chains. Those small dailies that were sold to large chains usually were acquired because they fit into one of the large company's geographic clusters of newspapers.

Local Control

A few owners of local dailies have sought to preserve future local control of their newspapers through devising employee-ownership strategies or by creating nonprofit foundations. Notable examples of employee ownership include the *Omaha World-Herald*, the *Milwaukee Journal Sentinel*, and the *Peoria Journal Star*. By 2003 the *Milwaukee Journal Sentinel* had grown from a single daily into a conglomerate known as Journal Communications that owned weeklies, broadcast stations and telecommunications and commercial-printing operating in 23 states. The company's directors concluded that the debt required to finance employee purchases of stock was becoming so burdensome that it threatened the company's ability to expand. Their solution was to dissolve the employee trust that controlled ownership, distribute super-voting shares to employees, and create a weak-voting class of shares to be listed on the New York Stock Exchange. Employees desiring to reduce their debt burden could convert their super-voting shares to the weak-voting shares in order to sell them in the public market. Control would thus be retained by the employees as long as enough of them retained super-voting shares.

Employee ownership turned out to be a trap for the *Peoria Journal Star*. In the early 1990's the owning family had envisioned a federally sanctioned Employee Stock Ownership Plan as a way to keep the paper out of chain hands. Unfortunately, after a dozen years, employees had become so wealthy on paper (by then they owned 91 percent of the stock), that they started cashing in their stock ahead of retirement. The company's directors realized that before long the company would not be able to finance buying back employee shares, and so they decided to sell the company to the Copley newspaper chain.

The model for creating a nonprofit foundation to retain local control of a daily was devised by the late Nelson Poynter, a passionate believer in the civic benefits of local ownership. His will left his ownership of the *St. Petersburg Times* to a foundation that created the Poynter institute, which provides mid-career training to journalists. Nackey Scripps Loeb, principal owner of the *Manchester* (NH) *Union Leader*, was no less passionate about the civic benefits of local ownership. When she died in 2000, she left her controlling ownership in the paper to the Nackey S. Loeb School of Communications, which provides training for aspiring journalists. Another approach to retaining local control was devised by Joe Smyth, owner of the *Delaware States News* in Dover and two other small dailies. In 1991 he created a nonprofit holding company to own his newspapers with the mission of keeping them independent and devoted to local coverage at the expense, if necessary, of high profit. More recently, the owner of the *Champaign* (IL) *News-Gazette* willed control of the paper to a nonprofit foundation that will support local charities. And the Ayers Family, owner of dailies in Anniston and Talladega, AL has announced that ownership of the papers will eventually transfer to a nonprofit foundation that will establish the Ayers Institute, which in collaboration with the University of Alabama will offer a graduate program in community journalism.

The Millennium

Noble as these various efforts are, it is unlikely that many other owning families will follow suit. The allure of the high prices paid by larger companies to acquire newspapers probably will be too great. Acquisition activity slowed down considerably in the early years of this century, owing chiefly to the recession that began in 2001, and has only slowly abated. The fact that newspaper deals dry up in a recession does not mean that newspapers suddenly are no longer valuable and desired by prospective purchasers. The problem is that potential sellers have a good idea of how much money they want, an amount usually expressed as a multiple of operating cash flow (earnings before non-cash charges like depreciation, interest and taxes). The going price for newspaper acquisitions has been a multiple of around 12 times operating cash flow. If earnings drop 15 to 20 percent in a recession, which is not unusual, the impact on how much an owner would get is large. It makes sense for owners to wait until business picks up substantially before selling. Thus as the economy improves we can expect a re-invigoration of acquisition activity. All of the pressures to buy and sell that were set out earlier in this chapter are still in place and assure continued ownership concentration of the nation's daily newspapers. How concentrated might newspaper ownership become? Dean Singleton, chief executive of MediaNews Group, a large newspaper chain, has opined that by the end of this century's first decade ten companies will control almost all of the nation's dailies (speech before California publishers, 1996). Given that 22 companies already control 70 percent of the nation's daily circulation, that may not be a far-fetched prediction.

Such ownership concentration raises important questions about the impact on the quality of public discourse. John C. Quinn, then a senior news executive for Gannett, in 1972 told those attending an International Press Institute conference that while increasing ownership concentration increases "the anxiety over evil," it also increases the capacity for good because a large company has the resources to invest in making newspapers better. He also said that "a publisher's instinct for good or evil is not determined by the number of newspapers he owns" (Compaine, 1980, p. 91). Taking note of Quinn's comments, Benjamin M. Compaine, executive director of the media at Harvard University's Program for Information Resources Policy, remarked that "the real argument hangs on the goodwill of the people in control . . . . a chain that is prone to milking its properties or throwing around its influence can infect numerous localities with poor or destructive journalism. The potential danger of group ownership lies in the concentration of financial, political and social power in relatively few people" (Compaine, 1979).

Clearly there is a potential for abuse, and one can look back in history to see what can happen. In the days of the press lords of the late 19th and early 20th Centuries, media companies had the power, and used it, to incite wars and riots. It was not unusual then for the owner of a chain of newspapers to dictate news coverage and editorial positions for the chain's newspapers. Fortunately, there were restrictions on the effects of this behavior, because in those days most towns of any size, and even many small ones, had more than one newspaper. Both newspapers in a town might slant the news, but each tended to slant it a different way. Today, though, almost all cities have but one newspaper; so the potential for mischief is greater. Still, today's newspapers are not the only sources of public information. Almost all towns now have radio stations, many have television stations, and, of course, there is the ubiquitous Internet. There are simply too many alternative sources of public information for any one media voice to get away with ignoring or distorting a major story for very long. Moreover, most modern newspaper chains practice what is called "local autonomy." Corporate headquarters may control local budgets carefully, including spending on news, and may direct improvements in methods of marketing and administration. But local direction of the news and editorial departments usually is left alone. However, observance of local autonomy pretty much depends on the good will and sophistication of the chief executives of large media corporations. Future corporate leaders might have different ideas.

So far, the buying of newspaper companies has with few exceptions involved primarily other newspapers companies. Buyers of newspaper properties generally have brought to their acquisitions at least some sensitivity to the special role that newspapers have in a democracy. Although more emphasis has been placed on profits in recent years, these companies mainly have not viewed their operations solely as economic machines, but but perceive them also as holders of a public trust. What might happen to a large newspaper company if its shareholders were to succumb to a lucrative offer from a non-newspaper company with no sensitivity to newspaper traditions or any concern about newspapers' public trust? Such a prospect may seem unlikely now, but minds can change.

REFERENCES

Audit Bureau of Circulations annual audits and biannual FAS-FAX reports.

Compaine, B. M. (1979). *Who owns the media? Concentration of ownership in the mass communications industry.* White Plains, NY: Knowledge Industry Publications, Inc.

Compaine, B.M. (1980), "The Newspaper Industry in the 1980s, An Assessment of Economics and Technology."White Plains, NY: Knowledge Industry Publications, Inc.

Dertouzos, J. N. (1978). Media conglomerates: Chains, groups and cross ownership. Presented to Federal Trade Commission Media Symposium. Washington, D.C.

Dirks, van Essen and Murray, press release, "Newspaper Acquisitions," January 2001, Vol. V, Issue 1, page l. *Editor and Publisher Yearbooks*, 1970-2004.

Emery, E. and Smith H. L. (1954). *The press and America.* New York: Prentice-Hall.

Morton Newspaper Newsletter, *Morton Research*, Inc. Monthly, 1976-2001.

Morton-Groves Newspaper Newsletter, Monthly, 2002-2004.
Mott, F. L. (1950). *American journalism.* New York: Macmillan.
Newspaper Association of America Fact Books, 1970-2003.
Project for Excellence in Journalism, The. (2004). The State of the News Media 2004. An Annual Report on American Journalism. Retrieved April 6, 2004, from http://www.stateofthemedia.org/narrative
Singleton, D. (1996). *Speech to California Publishers.*

INDEX

A

B

C

D

E

F

G

H

I

J

N

O

P

Q

R

S

T

U

V

W

Y

Z